ISBN 0-8373-6658-5

DANTES–58

Rudman's Question and Answers on the

DANTES SUBJECT STANDARDIZED TESTS

Subject Examination In . . .

ETHICS IN AMERICA

Questions and Answers

NLC

NATIONAL LEARNING CORPORATION
212 MICHAEL DRIVE, SYOSSET, NEW YORK 11791 (516) 921-8888

National Learning Corporation

212 Michael Drive, Syosset, New York 11791
(516) 921-8888
(800) 645-6337
Fax (516) 921-8743
www.passbooks.com

PRINTED IN THE UNITED STATES OF AMERICA

PASSBOOK®

NOTICE

NATIONAL LEARNING CORPORATION
212 Michael Drive
Syosset, New York 11791

Inquiries re licensing agreements should be addressed to:
The President
National Learning Corporation
212 Michael Drive
Syosset, New York 11791

PASSBOOK SERIES®

THE *PASSBOOK SERIES®* has been created to prepare applicants and candidates for the ultimate academic battlefield—the examination room.

At some time in our lives, each and every one of us may be required to take an examination—for validation, matriculation, admission, qualification, registration, certification, or licensure.

Based on the assumption that every applicant or candidate has met the basic formal educational standards, has taken the required number of courses, and read the necessary texts, the *PASSBOOK SERIES®* furnishes the one special preparation which may assure passing with confidence, instead of failing with insecurity. Examination questions—together with answers—are furnished as the basic vehicle for study so that the mysteries of the examination and its compounding difficulties may be eliminated or diminished by a sure method.

This book is meant to help you pass your examination provided that you qualify and are serious in your objective.

The entire field is reviewed through the huge store of content information which is succinctly presented through a provocative and challenging approach—the question-and-answer method.

A climate of success is established by furnishing the correct answers at the end of each test.

You soon learn to recognize types of questions, forms of questions, and patterns of questioning. You may even begin to anticipate expected outcomes.

You perceive that many questions are repeated or adapted so that you gain acute insights, which may enable you to score many sure points.

You learn how to confront new questions, or types of questions, and to attack them confidently and work out the correct answers.

You note objectives and emphases, and recognize pitfalls and dangers, so that you may make positive educational adjustments.

Moreover, you are kept fully informed in relation to new concepts, methods, practices, and directions in the field.

You discover that you are actually taking the examination all the time: you are preparing for the examination by "taking" an examination, not by reading extraneous and/or supererogatory textbooks.

In short, this PASSBOOK®, used directedly, should be an important factor in helping you to pass your test.

DANTES Subject Standardized Tests

INTRODUCTION

The DANTES (Defense Activity for Non-Traditional Education Support) subject standardized tests are comprehensive college and graduate level examinations given by the Armed Forces, colleges and graduate schools as end-of-subject course evaluation final examinations or to obtain college equivalency credits in the various subject areas tested.

Active military personnel may take CLEP (College-Level Examination Program) examinations without charge through the DANTES program. Information about DANTES and CLEP may be obtained from the Test Control Officers at military installations, and from Educational Testing Service in Princeton, New Jersey.

The DANTES Examination Program enables students to obtain college credit for what they have learned on the job, through self-study, personal interest, correspondence courses or by any other means. It is used by colleges and universities to award college credit to students who demonstrate that they know as much as students completing an equivalent college course. It is a cost-efficient, time-saving way for students to use their knowledge to accomplish their educational goals.

Most schools accept the American Council on Education (ACE) recommendations for the minimum score required and the amount of credit awarded, but not all schools do. Be sure to check the policy regarding the score level required for credit and the number of credits to be awarded.

Not all tests are accepted by all institutions. Even when a test is accepted by an institution, it may not be acceptable for every program at that institution. Before considering testing, ascertain the acceptability of a specific test for a particular course.

Colleges and universities that administer DANTES tests may administer them to any applicant -- or they may administer the tests only to students registered at their institution. Decisions about who will be allowed to test are made by the school. Students should contact the test center to determine current policies and schedules for DANTES testing.

Colleges and universities authorized to administer DANTES tests usually do so throughout the calendar year. Each school sets its own fee for test administration and establishes its own testing schedule. Contact the representative at the administering school directly to make arrangements for testing.

Your Experience Can Earn You Credit.

It's a fact. People learn in many ways . . . through hobbies . . . interests . . . reading . . . work. The DANTES Program is a way for you to "cash in" on what you've learned from *your* experiences.

Take advantage of DANTES tests to earn college credit for what you know. More than 30 subjects are offered. There are bound to be areas where your knowledge can pay off academically.

Here are just a few of the program's benefits for you:

- ***Time and Money Saver***
  You don't have to sit through classes and pay for courses when you already know most of what's being taught and can learn the rest yourself.
- ***Credits Widely Accepted***
  Hundreds of colleges nationwide now give credit for DANTES tests. And the list is growing!
- ***Quicker Route to the Degree***
  Your dreams are a little closer than you thought possible. With DANTES.
- ***Start Fast***
  No need to wait to take advanced courses. Passing DANTES tests gives you a real head start.

Sound Good? Here's How to Get Going!

Check with your college to see if it administers the DANTES program. If the college doesn't but would like to, ask that someone **call collect** (609-951-1756) or write to the DANTES Program Office (see address below).

If you find that your college *accepts* DANTES credits, but does not *administer* the tests, here's what you do —

- Write to the DANTES Program Office for a list of colleges that administer the test. Chances are there's one nearby.
- Call any of the ETS Regional Offices (Tucker, GA; Oakland, CA; Evanston, IL; Phoenix, AZ; and Princeton, NJ. DANTES tests are given at all of those locations.
- In the military or in the reserves? DoD employees and family members may be able to take DANTES tests at a Military Education Office or Navy Campus Office.

Check Out the DANTES Test List.

Match the titles with the things you have learned. Consider how much college credit you might gain by taking some DANTES tests.

Your college should have fact sheets with more information about each DANTES test. If not, please write to us directly, or e-mail us at dantes@chauncey.com. Tell us which subjects interest you.

DANTES Program Office
Educational Testing Service
Princeton, NJ 08541-0001

from the official announcement for instructional purposes

MATHEMATICS

Introductory College Algebra
- Functions and graphs
- Fundamental algebraic operations
- Linear equations, inequalities, word problems
- Radicals and exponents

Principles of Statistics
- Descriptive statistics
- Correlation, regression, probability
- Chance models, sampling, test of significance

SOCIAL SCIENCE

Art of the Western World
- Greece and Rome
- Romanesque and Gothic
- Renaissance
- Baroque, rococo, neoclassicism, romanticism
- Twentieth century

The Civil War and Reconstruction
- Causes of the war
- Political situations and the war between 1861 and 1865
- Reconstruction

Contemporary Western Europe
- Historical background
- Political systems
- Societal issues
- Integration

Criminal Justice
- Criminal behavior
- Police and criminal justice system
- Court system
- Corrections

Drug and Alcohol Abuse
- Pharmacological principles
- Classification of drugs
- Properties of alcohol, sedative hypnotics, narcotic analgesics, stimulants, hallucinogens and other drugs
- Prevention/treatment

Foundations of Education
- Contemporary issues in education
- Past and current influences on education
- Interrelationships between contemporary issues and influences, past or current, in education

Fundamentals of Counseling
- Counselor roles and functions
- Counseling relationships
- Theoretical approaches
- Social and cultural foundations

General Anthropology
- Anthropology branches, methodologies, and theoretical perspectives
- Physical anthropology and archeology
- Nature of culture
- Social, economic, political, and religious organization

A History of the Vietnam War
- The first Vietnam War (1945-1954)
- LBJ goes to war (1964-1965)
- Tet Offensive (1968)
- Vietnamizing the war (1968-1973)
- Peace is at hand (1968-1973)

Human/Cultural Geography
- Basic facts and concepts
- Culture and environment
- Spatial processes
- Regional geography

Introduction to Law Enforcement
- U.S. criminal justice and police systems
- Issues and organization of police
- Constitutional law and precedents

Introduction to the Modern Middle East
- Geography and early history
- Nineteenth and early twentieth century
- First and Second World wars
- Soviet-U.S. rivalry, Suez Crisis, Iranian Revolution, Persian Gulf Crisis

Lifespan Developmental Psychology
- Biological development
- Perception, learning, and memory
- Social, emotional, and personality development
- Cognition and language

BUSINESS

Business Law II
- Sales of goods
- Debtor and creditor relations
- Business organizations
- Property
- Commercial paper

Business Mathematics
- Basic computation — fractions, percents, ratios, graphs
- Business applications

Introduction to Business
- Economic, international, government, and ownership issues
- Management process
- Human resources and production
- Marketing, finance

Introduction to Computers with Programming in BASIC
- Hardware, software, data management
- Communications and connectivity
- BASIC: I/O, expressions, loops, conditionals, strings, arrays

Management Information Systems
- Systems theory, analysis and design of systems
- Hardware and software
- Database management
- Telecommunications
- Management of the MIS functional area, informational support

Money and Banking
- Role of commercial banks and other intermediaries
- Federal Reserve System
- Macroeconomic activity, U.S. policy, international system

Personal Finance
- Financial goals, budgeting, cash management, economic terminology
- Credit, debt, and major purchases
- Taxes and insurance
- Investments, retirement and estate planning

Personnel/Human Resource Management
- Employment actions
- Training and development
- Performance appraisals
- Compensation and personnel legislation
- Labor relations

Principles of Finance
- Financial statements and ratio analysis
- Capital budgeting
- Time value of money
- Break-even and leverage, common stock

Principles of Financial Accounting
- General concepts and principles
- Financial statements
- Accruals and deferrals
- Current accounts

Principles of Supervision
- Roles and responsibilities
- Management functions
- Legal issues, stress management, unions, productivity and quality concerns

Organizational Behavior
- Individual and group processes and characteristics
- Organizational processes and behavior
- Change and development processes

APPLIED TECHNOLOGY

Technical Writing
- Elements of technical reports
- Style, grammar, mechanics, organizing
- Types of technical reports

HUMANITIES

Ethics in America
- Ethical traditions
- Analysis of issues in relationships
- Tradition versus analysis of issues

Introduction to World Religions
- Dimensions and approaches
- Primal religions, Hinduism, Buddhism, Confucianism, Taoism, Judaism, Christianity, Islam

Principles of Public Speaking
- Topics and purposes
- Organization and supporting materials
- Language, style, and delivery
- Audience analysis and adaptation

PHYSICAL SCIENCE

Astronomy
- The solar system
- The sun and stars
- Our galaxy and the universe
- Measurement

Environment and Humanity: The Race to Save the Planet
- Ecological concepts
- Human impact on the environment
- Environmental management
- Political processes and the future

Here's to Your Health
- Mental health and behavior
- Human development
- Substance use and abuse
- Fitness and nutrition
- Risk factors and disease
- Consumer awareness, environmental concerns

Physical Geology
- Earth materials
- Surface processes
- Internal earth processes
- Applications

Principles of Physical Science I
- Physics: Newton's laws, energy, thermodynamics optics, electricity, and magnetism
- Chemistry: matter, atomic theory and structure, chemical reactions

NOTE:

MOST SCHOOLS DO NOT ACCEPT ALL TESTS FOR ALL PROGRAMS. CHECK WITH THE SCHOOL FROM WHICH YOU EXPECT TO RECEIVE CREDIT.

DANTES

Fact Sheet

Study Guide

Subject Standardized Tests

ETHICS IN AMERICA
SF 474

TEST INFORMATION

This test was developed to enable schools to award credit to students for knowledge equivalent to that which is learned by students taking the course. The test contains 111 items. The American Council on Education recommends that credit be awarded for a minimum of 63 correct responses, which is a scaled score of 46. This minimum credit-awarding score is equal to the mean score of students in the norming sample who received a grade of C in the course. However, some schools set their own standards for awarding credit and may require a higher score. Students should obtain this information from the institution where they expect to receive credit.

The essay will ask the examinee to analyze a morally problematic situation in terms of issues relevant to a decision and arguments for alternative positions; it will require the examinee to reach his or her own reasoned conclusion.

The essay section will not be scored by ETS. It will be forwarded to colleges and universities for their use in determining the award of credit.

CONTENT

The following topics, which are commonly taught in courses on this subject, are covered by this examination.

	Approximate Percent
I. Ethical traditions	29%

A. Greek views: Thucydides, Socrates, Plato, Aristotle
1. Relationship of morality to community
2. Duty to obey the law
3. Relationship between self-interest and virtue
4. Relationship between virtue and happiness
5. Conceptions of justice
6. Justice as the interest of the stronger
7. Justice as playing the proper role
8. Equity as proper ratio of benefits
9. Virtue and human nature
10. Nature and kinds of virtue
11. Virtue as excellence
12. Virtue, habit, and character
13. Voluntary action
14. Doctrine of the mean
15. The best human life

B. Biblical traditions: books of the law, prophets, gospels, writings of Paul
1. Relationship between law and morality
2. Role of the covenant
3. Role of the prophets in denouncing evil
4. Conceptions of justice in punishment and in the distribution of wealth
5. Relationship between morality and community
6. Place of mercy and love
7. Apparent paradoxes in the traditions

C. Moral law: Epictetus, Aquinas, Hobbes, Locke, Jefferson, Kant, Rawls, King, Rousseau, and others
1. Views of human nature
2. Basis of natural law
3. Natural rights and duties
4. Relationship of natural law to religious belief

Approximate Percent

5. Source of political authority
6. Social contract
7. Relationship of natural law and human law
8. Justifiability of revolution or civil disobedience
9. Equality and liberty
10. Deontological vs. consequential moral frameworks
11. Hypothetical and categorical imperatives
12. Good will and duty for its own sake
13. Persons as ends in themselves, not as means only
14. Loyalty and virtue
15. Relationship of morality to self-interest and happiness
16. Conceptions of justice
17. Fairness and distributive justice

D. Consequential ethics: Epicurus, Bentham, Mill
1. Consequential vs. deontological moral frameworks
2. Hedonism and kinds of pleasure
3. Self-interest or prudence and the common good
4. Principle of utility or greatest happiness principle
5. Felicific (or hedonic) calculus
6. Conscience and external sanctions of morality
7. Self-regarding and other-regarding conduct
8. Individual liberty and the limits of the state's authority

E. Feminist ethics
1. Critique of standard ethics
2. Alternative visions - care ethic and ethic of trust

II. Ethical analysis of issues arising in interpersonal and personal-societal relationships and in professional and occupational roles (such as law, government, medicine, business military service, journalism), issues such as but not limited to — 35%

A. Doing good, not doing harm, and preventing harm
B. Obligations to strangers
C. Special moral relationships (such as parenthood)
D. Conflict of duties
E. Autonomy and privacy
F. Fidelity, accountability, and trustworthiness
G. Veracity
H. Right to information
I. Confidentiality
J. Informed consent
K. Institutional responsibilities
L. Professional code of ethics
M. Paternalism
N. Retributive and distributive justice
O. Balancing harms

III. Relationships between ethical traditions and the ethical analysis of situations — 36%

SAMPLE QUESTIONS

1. Which of the following thinkers evaluates the morality of actions in terms of the amount of pleasure and pain they produce?

   (A) Aristotle
   (B) Kant
   (C) Bentham
   (D) King

2. According to which of the following philosophers is it immoral to commit suicide?

   (A) The hedonist Epicurus, because suicide decreases one's pleasure
   (B) The Stoic Epictetus, because suicide is contrary to nature
   (C) Kant, because persons who commit suicide treat themselves as a means only
   (D) Mill, because suicide never maximizes utility

3. For a utilitarian, the ethical evaluation of a decision to tell a lie will NOT depend on whether the

   (A) person to whom the lie is told will benefit from the lie
   (B) person telling the lie would mind someone else lying to him or her
   (C) person telling the lie will suffer if the truth is told
   (D) lie will cause suffering to third parties

Directions: The following questions are based on fictional situations and are in a different format from that of other questions in this test. Moral relevance and the moral reasons are to be understood in terms of the theories presented in the course materials. Judge a consideration to be morally relevant if at least one of those views would allow its relevance.

Preceding each group of numbered phrases or statements is a question that can be answered "Yes" or "No." Answer this question separately for each phrase or statement in the group. Do not assume that there are more "Yes" answers or more "No" answers.

Questions 4-10 are based on the following situation.

Dr. DePaolo, chief of psychiatry at University Hospital, submitted a research plan to the hospital's Institutional Review Board to use patients in the hospital to test the effectiveness of a new type of drug for treatment of severe depression. The doctor proposed a "double-blind" design in which subjects would be randomly assigned to one of three groups: one group to receive the experimental drug, a second group to receive a standard agent for treatment of depression, and a third group to receive a placebo (a harmless inert substance). Dr. DePaolo as the experimenter was not to know which subjects were assigned to each group. The plan was approved. A third of the way through the experimental trials, one of the subjects complained to Dr. DePaolo, "Doctor, since I have been taking this medication, I have been feeling much more depressed and have been thinking again about committing suicide."

Is the following an issue that would have been relevant for a member of the Institutional Review Board to raise in judging the plan from the point of view of the ethics of research on human subjects?

4. Whether the patients were depressed to a degree that would impair their judgment in agreeing to participate in the trial

5. What information was to be given to the subjects about the possible risks and benefits of the trial to them

6. Whether the hospital would incur unreimbursed costs by participating in the trial

Is the following an appropriate course of action for Dr. DePaolo to follow concerning the depressed subject if Dr. DePaolo applies the principle "First, do not harm"?

7. Record the comment as data for the experimental trial and continue the trial

8. Wait until the end of the trial to treat the subject with the drug that has been found to be most effective for that subject

9. Remove the subject from the experimental trial and resume treatment with an approved drug suitable for that subject

10. Unseal the trial records to determine which experimental group the subject is in before deciding whether to continue the trial for that subject

ESSAY QUESTION

Choose one of the following situations to analyze.

Situation: While off duty, Timmons, a police officer, finds a package containing $1,000 on the floor of a supermarket. The package and its contents give no indication of its owner's name. In the package, however, is a betting ticket for a legal lottery which the owner of the package could use to identify the package if it is reported found. Timmons needs cash in order to make a down payment on a new house. Without new housing, Timmons's children will continue to be exposed to high levels of radon, a dangerous gas emanating from the granite underneath Timmons's present house. Radon exposure, if prolonged, can cause cancer.

Decision: Should Timmons keep the package and its contents without reporting it to the store or the local authorities?

Part 1. List the most significant moral principles or issues and the most significant facts, including any undecided factual issues, that have a bearing on making the decision, and describe the alternatives available.

Part 2. Then, using this analysis, take an ethical position about what decision should be made, and support your position with the best ethical arguments you can make.

Part 3. Then, in the most reasonable way that you can, state a view that is strongly opposed to your position in the most reasonable way that you can, and make the best ethical arguments you can for this opposing view.

Part 4. Then reconsider your position, and say whether and why it should be changed or modified in view of the opposing arguments you have made.

Write one and no more than two paragraphs on each part.

Allocate your time equally among the four parts.

REFERENCE PUBLICATIONS

This test is based on a variety of textbooks and materials, as well as on the Annenberg/CPB video course *Ethics in America* and its accompanying study guide.

Barcalow, Emmett. *Moral Philosophy Theory and Issues*. Belmont, CA: Wadsworth, 1994. An interpretive analysis without primary source references.

Goldberg, David T. *Ethical Theory & Social Issues: Historical Texts & Contemporary Readings*. Fort Worth, TX: Harcourt Brace College, 1989. Anthology of classical readings in ethics. Articles on affirmative action; censorship and pornography; sexual morality and AIDS; and other applied ethics topics.

Newton, Lisa H., ed. *Ethics in America Source Reader*. Englewood Cliffs, NJ: Prentice-Hall, Inc., 1989.

Sommers, Christina, and Fred Sommers. *Vice and Virtue in Everyday Life*. 3d ed. Fort Worth, TX: Harcourt Brace College, 1993. Short selections, both historical and contemporary, in ethical theory and its applications.

White, James E., ed. *Contemporary Moral Problems*. 4th ed. Minneapolis/St. Paul, MN: West Publishing, 1994. Anthology of readings on contemporary moral issues with an opening section covering ethical theory. Includes feminist theory and selections from feminist writers.

CREDIT RECOMMENDATIONS

The Center For Adult Learning and Educational Credentials of the American Council on Education (ACE) has reviewed and evaluated the DANTES examination development process. The American Council on Education has made the following recommendations:

Area or Course Equivalent:	Ethics in America
Level:	Lower level Baccalaureate
Amount of Credit:	Three (3) semester hours
Minimum Credit-awarding Score:	Scaled score of 46 (63 correct responses)
Source:	ACE Commission on Educational Credit and Credentials

TECHNICAL INFORMATION

Twenty four colleges and universities participated in the norming during the fall of 1994. These schools ranged from large state universities to small private schools. Geographic distribution was wide, with schools from all parts of the United States involved. A total of 592 students who were completing or had recently completed an elementary course in this subject took part in the norming.

Raw scores were converted to a standard score scale with a mean of 50 and a standard deviation of 10 using the formula $S = AX + B$. The A and B parameters were generated from a mean of 67.8598 and a standard deviation of 11.9541. The parameters are $A = 0.8365$ and $B = -6.7647$.

NORMING STATISTICS

(based on 590 students)

Number of items	111
Mean (Raw Score)	67.9
Median (Raw Score)	66.6
Standard deviation (Raw Score)	12.0
KR-20 Reliability	0.85
Standard Error of measurement	
(Raw Score)	4.6
(Scaled Score)	3.9

Average Grade/Test Score Correlation, based on a sample of 546 students at 22 institutions (an average of within-school correlations, weighted by the number of students tested in each school):	.46
Average scaled score for students whose teachers gave them a grade of A for the course:	54
Average scaled score for students whose teachers gave them a grade of B for the course:	50
Average scaled score for students whose teachers gave them a grade of C for the course:	46

Colleges and universities that would like to have review copies of tests, additional information about the national norming, or assistance in local norming or score validation studies should write to: DANTES Program, Mail Stop 31-X, Educational Testing Service, Princeton, New Jersey 08541-0001.

It is advisable that schools develop a consistent policy about awarding credit based on scores from this test and that the policy be reviewed periodically. ETS will be happy to help schools in this effort.

Correct responses to sample questions: 1.C; 2.C; 3.B. PART II: 4.A; 5.A; 6.B; 7.B; 8.B; 9.A; 10.B.

HOW TO TAKE A TEST

You have studied hard, long, and conscientiously.

With your official admission card in hand, and your heart pounding, you have been admitted to the examination room.

You note that there are several hundred other applicants in the examination room waiting to take the same test.

They all appear to be equally well prepared.

You know that nothing but your best effort will suffice. The "moment of truth" is at hand: you now have to demonstrate objectively, in writing, your knowledge of content and your understanding of subject matter.

You are fighting the most important battle of your life—to pass and/or score high on an examination which will determine your career and provide the economic basis for your livelihood.

What extra, special things should you know and should you do in taking the examination?

BEFORE THE TEST

YOUR PHYSICAL CONDITION IS IMPORTANT

If you are not well, you can't do your best work on tests. If you are half asleep, you can't do your best either. Here are some tips:

1. Get about the same amount of sleep you usually get. Don't stay up all night before the test, either partying or worrying—DON'T DO IT.
2. If you wear glasses, be sure to wear them when you go to take the test. This goes for hearing aids, too.
3. If you have any physical problems that may keep you from doing your best, be sure to tell the person giving the test. If you are sick or in poor health, you really cannot do your best on any test. You can always come back and take the test some other time.

AT THE TEST

EXAMINATION TECHNIQUES

1. Read the *general* instructions carefully. These are usually printed on the first page of the examination booklet. As a rule, these instructions refer to the timing of the examination; the fact that you should not start work until the signal and must stop work at a signal, etc. If there are any *special* instructions, such as a choice of questions to be answered, make sure that you note this instruction carefully.

2. When you are ready to start work on the examination, that is as soon as the signal has been given, read the instructions to each question booklet, underline any key words or phrases, such as *least, best, outline, describe*, and the like. In this way you will tend to answer as requested rather than discover on reviewing your paper that you *listed without describing,* that you selected the *worst* choice rather than the *best* choice, etc.

3. If the examination is of the objective or so-called multiple-choice type, that is, each question will also give a series of possible answers: A, B, C, or D, and you are called upon to select the best answer and write the letter next to that answer on your answer paper, it is advisable to start answering each question in turn. There may be anywhere from 50 to 100 such questions in the three or four hours allotted and you can see how much time would be taken if you read through all the questions before beginning to answer any. Furthermore, if you come across a question or a group of questions which you know would be difficult to answer, it would undoubtedly affect your handling of all the other questions.

4. If the examination is of the essay-type and contains but a few questions, it is a moot point as to whether you should read all the questions before starting to answer any one. Of course if you are given a choice, say five out of seven and the like, then it is essential to read all the questions so you can eliminate the two which are most difficult. If, however, you are asked to answer all the questions, there may be danger in trying to answer the easiest one first because you may find that you will spend too much time on it. The best technique is to answer the first question, then proceed to the second, etc.

5. Time your answers. Before the examination begins, write down the time it started, then add the time allowed for the examination and write down the time it must be completed, then divide the time available somewhat as follows:

   a. If 3 ½ hours are allowed, that would be 210 minutes. If you have 80 objective-type questions, that would be an average of about 2 ½ minutes per question. Allow yourself no more than 2 minutes per question, or a total of 160 minutes, which will permit about 50 minutes to review.

   b. If for the time allotment of 210 minutes, there are 7 essay questions to answer, that would average about 30 minutes a question. Give yourself only 25 minutes per question so that you have about 35 minutes to review.

6. The most important instruction is *to read each question* and make sure you know what is wanted. The second most important instruction is to *time yourself properly* so that you answer every question. The third most important instruction is to *answer every question.* Guess if you have to but include something for each question, Remember that you will receive no credit for a blank and will probably receive some credit if you write something in answer to an essay question. If you guess a letter, say "B" for a multiple-choice question, you may have guessed right. If you leave a blank as the answer to a multiple-choice question, the examiners may respect your feelings but it will not add a point to your score. Some exams may penalize you for wrong answers, so in such cases *only*, you may not want to guess unless you have some basis for your answer.

7. Suggestions

   a. Objective-Type Questions

      (1) Examine the question booklet for proper sequence of pages and questions.

      (2) Read all instructions carefully.

      (3) Skip any question which seems too difficult; return to it after all other questions have been answered.

      (4) Apportion your time properly; do not spend too much time on any single question or group of questions.

      (5) Note and underline key words—*all, most, fewest, least, best, worst, same, opposite*.

      (6) Pay particular attention to negatives.

      (7) Note unusual option, e.g., unduly long, short, complex, different or similar in content to the body of the question.

      (8) Observe the use of "hedging" words—*probably, may, most likely, etc.*

      (9) Make sure that your answer is put next to the same number as the question.

      (10) Do not second guess unless you have good reason to believe the second answer is definitely more correct.

      (11) Cross out original answer if you decide another answer is more accurate; do not erase, *until* you are ready to hand your paper in.

      (12) Answer all questions; guess unless instructed otherwise.

      (13) Leave time for review.

b. Essay-Type Questions

(1) Read each question carefully.

(2) Determine exactly what is wanted. Underline key words or phrases.

(3) Decide on outline or paragraph answer.

(4) Include many different points and elements unless asked to develop any one or two points or elements.

(5) Show impartiality by giving pros and cons unless directed to select one side only.

(6) Make and write down any assumptions you find necessary to answer the question.

(7) Watch your English, grammar, punctuation, choice of words.

(8) Time your answers; don't crowd material.

8. Answering the Essay Question

Most essay questions can be answered by framing the specific response around several key words or ideas. Here are a few such key words or ideas:

M's: manpower, materials, methods, money, management

P's: purpose, program, policy, plan, procedure, practice, problems, pitfalls, personnel, public relations

a. Six basic steps in handling problems:

(1) preliminary plan and background development

(2) collect information, data and facts

(3) analyze and interpret information, data and facts

(4) analyze and develop solutions as well as make recommendations

(5) prepare report and sell recommendations

(6) install recommendations and follow up effectiveness

b. Pitfalls to Avoid

(1) *Taking Things for Granted*
A statement of the situation does not necessarily imply that each of the elements is necessarily true; for example, a complaint may be invalid and biased so that all that can be taken for granted is that a complaint has been registered

(2) *Considering only one side of a situation*
Wherever possible, indicate several alternatives and then point out the reasons you selected the best one.

(3) *Failing to indicate follow up*
Whenever your answer indicates action on your part, make certain that you will take proper follow-up action to see how successful your recommendations, procedures, or actions turn out to be.

(4) *Taking too long in answering any single question*
Remember to time your answers properly.

EXAMINATION SECTION

EXAMINATION SECTION

TEST 1

DIRECTIONS: Each question or incomplete statement is followed by several suggested answers or completions. Select the one that BEST answers the question or completes the statement. *PRINT THE LETTER OF THE CORRECT ANSWER IN THE SPACE AT THE RIGHT.*

1. A *dogmatic* person is someone who 1.___
   I. listens to others' arguments and then decides
   II. listens to others' arguments but never decides
   III. never listens to others' arguments
   IV. offers reasons for his view
   V. never offers reasons for his view

   The CORRECT answer is:
   A. I *only*
   B. II *only*
   C. III, IV
   D. IV *only*
   E. III, V

2. An *eclectic* philosophy is one that 2.___
   I. draws upon a variety of theories for its conclusion
   II. relies upon a fundamental component for its basis
   III. speculates freely and relies upon nothing
   IV. combines several elements
   V. has a unitary simplicity

   The CORRECT answer is:
   A. I, IV
   B. I, V
   C. II, III
   D. II, III, IV
   E. III, IV, V

3. The psychological theory which has provided a rationale for capitalism is 3.___
   A. altruism
   B. egoism
   C. relativism
   D. Freudianism
   E. none of the above

4. An EQUIVOCATION is a logical fallacy caused by 4.___
   A. false assumptions
   B. invalid inferences
   C. ambiguous statements
   D. listeners jumping to conclusions
   E. all of the above

5. The study of ESCHATOLOGY concerns ____ things. 5.___
   A. first
   B. last
   C. all
   D. God's
   E. human

6. The ancient Greek root of the English word "ethics" meant 6.___
   A. life
   B. virtue
   C. love
   D. duty
   E. custom

7. EUPHEMISMS are words that 7.___
I. don't mean anything
II. can mean just about anything
III. substitute for other, more concrete, words
IV. substitute for other, more vague, words
V. should be used sparingly, if at all

The CORRECT answer is:
A. I, II
B. II *only*
C. II, III, V
D. III, V
E. IV, V

8. Some existentialist philosophers are 8.___
I. Hume
II. Marx
III. Kierkegaard
IV. Sartre
V. Camus

The CORRECT answer is:
A. I, II
B. I, II, III
C. I, III, IV
D. II, IV, V
E. III, IV, V

9. EXISTENTIALISM maintains that 9.___
I. I am, therefore I think
II. no one knows anything
III. life is absurd
IV. suicide is noble
V. we don't choose to be born

The CORRECT answer is:
A. I, III, IV, V
B. II, III
C. II, IV, V
D. III, IV, V
E. all of the above

10. Facts are those things *contrasted* with 10.___
I. a priori truths
II. logical axioms
III. necessary truths
IV. falsehoods
V. speculations

The CORRECT answer is:
A. I *only*
B. I, II
C. I, II, III
D. II, III, V
E. III, IV, V

11. FATALISM is the view that 11.___
I. the Fates know what is in store for us
II. you will not die until your time is come
III. everything that happens was meant to happen
IV. nothing that happens was meant to happen
V. no one knows what it means

The CORRECT answer is:
A. I, II
B. II, III
C. II, IV
D. II, IV, V
E. III, V

12. Philosophers who believed in the "Greatest Happiness Principle" were 12.___
I. Plato
II. Bentham
III. Mill
IV. Rousseau
V. Epicurus

The CORRECT answer is:
A. I, II, III, V
B. I, III, IV, V
C. II, III, IV
D. II, III, IV, V
E. all of the above

13. *Deontological* ethics emphasizes the concept(s) of 13.___
I. justice
II. rights
III. duty
IV. goodness
V. virtue

The CORRECT answer is:
A. I *only*
B. II *only*
C. II, III
D. II, III, IV
E. III, IV, V

14. *Teleological* ethics emphasizes 14.___
A. justice
B. rights
C. duty
D. goodness
E. virtue

15. Among philosophers emphasizing *hedonism* are 15.___
I. Plato
II. Epicurus
III. Kant
IV. Hobbes
V. Bentham

The CORRECT answer is:
A. I, II
B. I, III
C. II, III, IV
D. II, IV, V
E. II, V

16. Hegel's philosophy was a form of 16.___
I. materialism
II. idealism
III. dialecticism
IV. absolutism
V. opportunism

The CORRECT answer is:
A. I, II
B. II, III
C. II, III, IV
D. II, IV
E. III, IV, V

17. Philosophers who were *heavily* influenced by Hegel were 17.___
I. Aristotle
II. Plotinus
III. Marx
IV. Fichte
V. Kierkegaard

The CORRECT answer is:
A. I, III
B. II, III
C. II, III, IV
D. III, IV
E. III, V

18. The view that reality changes constantly and that no one can "step into the same river twice" is attributed to 18.___
   A. Plato
   B. Aristotle
   C. Plotinus
   D. Heraclitus
   E. Democritus

19. The first historians are *commonly* believed to be 19.___
   I. Marcus Aurelius
   II. Aristotle
   III. Goebbels
   IV. Thucydides
   V. Herodotus

   The CORRECT answer is:
   A. I, II
   B. I, II, III
   C. II, III
   D. II, IV, V
   E. IV, V

20. The author of THE SOCIAL CONTRACT was 20.___
   A. Hobbes
   B. Locke
   C. Rousseau
   D. Hegel
   E. Marx

21. The author of A TREATISE OF HUMAN NATURE is 21.___
   A. Locke
   B. Hobbes
   C. Rousseau
   D. Hume
   E. Berkeley

22. A *hypothetical imperative* claims that 22.___
   A. everything we want is only for something else
   B. life is tentative and therefore hypothetical
   C. if you want one thing, you must do another
   D. some things are desired for themselves
   E. none of the above

23. The law of identity 23.___
   I. cannot be false
   II. can be true
   III. must be true
   IV. is an a priori truth
   V. all of the above

   The CORRECT answer is:
   A. I, III, IV
   B. I, IV
   C. II, IV
   D. III *only*
   E. V *only*

24. The Enlightenment was characterized by 24.___
   A. love, emotion and faith
   B. reason, logic and math
   C. capitalism
   D. feudalism
   E. none of the above

25. The Renaissance was characterized by 25.___
   I. philosophical sophistication
   II. artistic brilliance
   III. specialization
   IV. dilettantes
   V. generalizing

The CORRECT answer is:
A. I *only*
B. I, II, III
C. II, IV
D. II, III, IV
E. III, IV, V

26. One of Germany's MOST famous rationalists is 26.___
A. Nietzsche
B. Schopenhauer
C. Schilling
D. Scheler
E. Kant

27. LIBERTARIANISM relies upon the views of 27.___
A. Marx
B. Locke
C. Aristotle
D. Hobbes
E. Machiavelli

28. John Locke is noted for emphasizing 28.___
A. inalienable rights
B. property rights
C. labor as allowing property appropriation
D. life, liberty, and the protection of property
E. all of the above

29. METAPHYSICS is MOST intimately associated with 29.___
A. ethics
B. epistemology
C. ontology
D. aesthetics
E. logic

30. The author of the MONODOLOGY is 30.___
A. Descartes
B. Kant
C. Maimonides
D. Leibniz
E. Fichte

31. EMOTIVISM is the theory that says 31.___
A. everyone believes something different about morality
B. people are too emotional to do philosophy
C. if something makes you unhappy, it is wrong
D. if something makes you happy, it is right
E. saying something is right means that you like it

32. NEUTRAL MONISM was believed by 32.___
A. G. E. Moore
B. William James
C. Bertrand Russell
D. no one
E. ancient Greeks

33. A NON SEQUITUR is a statement that 33.___
I. follows logically from some premises
II. does not follow logically from some premises
III. makes sense given other statements conjoined with it
IV. does not make sense
V. makes people laugh frequently

The CORRECT answer is:
A. I, III
B. II, IV, V
C. III, V
D. IV *only*
E. V *only*

34. If God is *omniscient*, it means He is 34.___
A. all-powerful
B. all-knowing
C. ever-present
D. eternal
E. all of the above

35. If God is *omnipresent*, it means He is 35.___
A. all-powerful
B. all-knowing
C. ever-present
D. eternal
E. all of the above

36. If God is *omnipotent*, it means He is 36.___
A. all-powerful
B. all-knowing
C. ever-present
D. eternal
E. all of the above

37. The view that claims that emotions are philosophically valid is 37.___
A. materialism
B. idealism
C. rationalism
D. existentialism
E. none of the above

38. The statement "I am lying" is paradoxical because it is 38.___
I. true if it is false
II. false if it is true
III. false
IV. reflexive
V. none of the above

The CORRECT answer is:
A. I *only*
B. I, II
C. II *only*
D. III, IV
E. V *only*

39. Some of the MOST famous American philosophers are 39.___
I. Rousseau
II. Locke
III. Peirce
IV. Dewey
V. James

The CORRECT answer is:
A. I, II
B. I, III
C. I, III, IV
D. II, IV, V
E. III, IV, V

40. Among the contemporary European philosophies are 40.___
I. existentialism
II. structuralism
III. phenomenology
IV. Marxism
V. none of the above

The CORRECT answer is:
A. I, II, III, IV
B. I, III, IV
C. II, III, IV
D. II, IV
E. V *only*

41. PRAGMATISM is a philosophy that claims that 41.___
I. one should only do what is necessary
II. consequences of actions ought to be considered when evaluating a theory
III. empiricism is the best criterion of truth
IV. idealism is to be rejected
V. all of the above

The CORRECT answer is:
A. I *only*
B. I, II, III
C. II, III, IV
D. III, IV
E. V *only*

42. Famous pragmatists are 42.___
I. Marx
II. Hegel
III. Dewey
IV. Peirce
V. all of the above

The CORRECT answer is:
A. I, II
B. I, III
C. II, III, IV
D. III, IV
E. V *only*

43. Probability is an *essential* feature of 43.___
I. a priori reasoning
II. a posteriori reasoning
III. science
IV. mathematics
V. all of the above

The CORRECT answer is:
A. I, II, III
B. I, III
C. II, III
D. III, IV
E. V *only*

44. The *first* philosopher to discuss psychology was 44.___
A. Plato
B. Aristotle
C. Plotinus
D. Freud
E. Bacon

45. PHILOSOPHICAL REALISM states that 45.___
A. things in the world exist independently of humans
B. the things we get to know are eternal
C. objects are those things seen by human beings
D. what we know will help us
E. what we don't know won't hurt us

46. Bertrand Russell, the famous British philosopher, FIRST became famous for his 46.___
A. aristocratic background
B. rich wife
C. radical ideas
D. mathematical system
E. none of the above

47. The famous dictum, "All property is theft," was written by the French philosopher 47.___
A. Rousseau
B. Montaigne
C. Saint-Simon
D. Comte
E. none of the above

48. Philosophers who were skeptics include 48.___
I. Aristotle
II. Descartes
III. Hume
IV. Reid
V. none of the above

The CORRECT answer is:
A. I, II
B. II, III
C. II, III, IV
D. III, IV
E. V

49. Semantics concerns the 49.___
A. tower of Babel
B. structure of language
C. meaning of terms
D. vagueness of conversational language
E. all of the above

50. A statement is a(n) 50.___
A. sentence that can be proven empirically
B. sentence that expresses a thought
C. sentence that can be true or false
D. sentence that cannot be true or false
E. exclamation

KEY (CORRECT ANSWERS)

1. E	11. B	21. D	31. E	41. C
2. A	12. A	22. C	32. C	42. D
3. B	13. C	23. A	33. B	43. C
4. C	14. D	24. B	34. B	44. B
5. B	15. E	25. C	35. C	45. A
6. E	16. C	26. E	36. A	46. D
7. D	17. E	27. B	37. D	47. C
8. E	18. D	28. E	38. B	48. B
9. A	19. E	29. C	39. E	49. C
10. C	20. C	30. D	40. A	50. C

TEST 2

DIRECTIONS: Each question or incomplete statement is followed by several suggested answers or completions. Select the one that BEST answers the question or completes the statement. *PRINT THE LETTER OF THE CORRECT ANSWER IN THE SPACE AT THE RIGHT.*

1. SOLIPSISM is a view that claims that 1.___
   A. everyone else is a figment of my imagination
   B. only I exist
   C. my own existence is the only one I can prove
   D. probably other people exist
   E. all of the above

2. The SOCRATIC METHOD *primarily* consists of 2.___
   A. antagonizing others by preaching at them
   B. asking questions
   C. asking atheistic questions
   D. talking to Plato
   E. writing dialogues

3. SOPHISTRY is 3.___
   A. the art of language manipulation
   B. the practice of teaching for money
   C. the art of cross-examination
   D. knowing many things
   E. pretending to be ignorant

4. STATE-OF-NATURE theory is another name for 4.___
   A. utilitarianism
   B. socialism
   C. capitalism
   D. utopianism
   E. contract theory

5. A STOIC is a person who believes that 5.___
   A. only through suffering can we learn anything
   B. pleasures of the body are evil
   C. happiness comes primarily through knowledge
   D. ignorance is bliss
   E. none of the above

6. SUBJECTIVE IDEALISM claims that 6.___
   I. a person's own standards are what ought to guide him/her
   II. our experience of the real world is what gives us knowledge
   III. the entire world exists primarily through our mind
   IV. to perceive something is to know it
   V. all of the above

   The CORRECT answer is:
   A. I *only*
   B. I, II
   C. II, III
   D. III, IV
   E. V *only*

7. SUBSTANCE is one of the *primary* concepts in 7.___
A. epistemology B. aesthetics
C. political philosophy D. metaphysics
E. theology

8. The "sunnum bonum," a frequently used concept in ethics, means the 8.___
A. happiness of the greatest number
B. optimal efficiency for the whole society
C. most pleasure for the individual involved
D. highest good
E. all of the above

9. A *teleological* theory is one that emphasizes 9.___
I. motives II. actions
III. consequences IV. natural propensities
V. purposes

The CORRECT answer is:
A. I, II B. I, II, III
C. II, III, IV D. II, IV, V
E. IV, V

10. "Bachelors are unmarried men" is an example of a(n) 10.___
A. irrelevant claim B. insult to bachelors
C. tautology D. contradiction
E. none of the above

11. 1. All pigs are green. 11.___
2. Harvey is a pig.
3. Therefore, Harvey is green.
The preceding syllogism is
A. valid B. sound
C. neither valid nor sound D. both valid and sound
E. valid and unsound

12. Something has *intrinsic* value if it has 12.___
A. an emotional connotation for the individual
B. value in itself
C. value in the consequences it brings
D. value according to society
E. none of the above

13. Something has *extrinsic* value if it has 13.___
A. an emotional connotation for the individual
B. value in itself
C. value in the consequences it brings
D. value according to society
E. none of the above

14. The philosopher responsible for claiming that this is the "best of all possible worlds" is 14.___
A. Aristotle B. Plato
C. Descartes D. Leibniz
E. St. Anselm

15. The philosopher and dramatist who *satirized* the view of the "best of all possible worlds" was 15.___
A. Voltaire
B. Euripides
C. Hume
D. Sophocles
E. Shakespeare

16. A *weltanschauung* is a 16.___
A. bizarre perspective on life
B. philosophical position requiring scholarly research
C. German dessert
D. worldview
E. none of the above

17. The philosopher who became well-known for discussing "language games" is 17.___
A. Bertrand Russell
B. David Hume
C. G. E. Moore
D. Herbert Spencer
E. Ludwig Wittgenstein

18. BEING AND NOTHINGNESS was written by 18.___
A. Camus
B. Heidegger
C. Sartre
D. Wittgenstein
E. none of the above

19. An advocate of sociobiology is 19.___
A. Bertrand Russell
B. Ludwig Wittgenstein
C. Edmund Wilson
D. Jean-Paul Sartre
E. Eric Fromm

20. The famous, though short, book THE MEDITATIONS, was written by 20.___
A. Pascal
B. Descartes
C. Hume
D. Camus
E. none of the above

21. An advocate of free will who was OPPOSED to determinism was 21.___
A. Skinner
B. Freud
C. Sartre
D. Hume
E. none of the above

22. An advocate of determinism who was OPPOSED to free will was 22.___
A. Skinner
B. Freud
C. Sartre
D. Hume
E. none of the above

23. The NICHOMACHEAN ETHICS was written by 23.___
A. Plato
B. Aristotle
C. Descartes
D. Plotinus
E. none of the above

24. BEYOND GOOD AND EVIL was written by 24.___
A. Kierkegaard
B. Descartes
C. Russell
D. Nietzsche
E. Schopenhauer

25. The book ON LIBERTY, very influential in early American legal theory, was written by 25.___
I. Jeremy Bentham
II. David Hume
III. John Locke
IV. John Stuart Mill
V. John Stuart Mill and Harriet Taylor

The CORRECT answer is:
A. I *only*
B. II *only*
C. III, IV
D. III, IV, V
E. IV, V

26. A behaviorist famous for his experiments with pigeons was 26.___
A. Watson
B. Pavlov
C. Skinner
D. Freud
E. none of the above

27. A behaviorist famous for his experiments with dogs was 27.___
A. Watson
B. Pavlov
C. Skinner
D. Freud
E. none of the above

28. A fundamental advocate of LOGICAL POSITIVISM was 28.___
A. Ludwig Wittgenstein
B. David Hume
C. John Dewey
D. William James
E. A. J. Ayer

29. The American philosopher *most* associated with EMOTIVISM was 29.___
A. G. E. Moore
B. William James
C. John Dewey
D. C. L. Stevenson
E. none of the above

30. Isaac Newton is well-known for his works on 30.___
A. philosophy
B. mathematics
C. astrology
D. physics
E. all of the above

31. Philosophical reasoning carefully AVOIDS 31.___
A. superstition
B. supernatural entities
C. emotion
D. fallacies
E. all of the above

32. The ancient Greek philosopher who claimed that the primordial element of the universe was water was 32.___
A. Aristotle
B. Epictetus
C. Socrates
D. Plato
E. Thales

33. Socrates ended his life by 33.___
I. hanging himself
II. committing suicide
III. being condemned by the state
IV. drinking hemlock
V. being thrown to lions

The CORRECT answer is:
A. I, II
B. II, III
C. III, IV
D. III, V
E. V *only*

34. Socrates likened himself to a 34.___
A. god
B. humble peasant
C. noble lord
D. gadfly
E. none of the above

35. The subjects Socrates was interested in were 35.___
I. friends and pleasure
II. talking and drinking
III. truth and beauty
IV. justice and virtue
V. all of the above

The CORRECT answer is:
A. I, II, III
B. III, IV
C. IV *only*
D. V *only*
E. none of the above

36. According to Hobbes, no man can be required to 36.___
A. obey the government
B. keep his promises
C. act self-destructively
D. share his goods with others
E. all of the above

37. Hobbes claimed that justice can be obtained 37.___
A. only through benevolence
B. only with loved ones
C. only in civil society
D. only in the state of nature
E. never

38. TWO TREATISES ON CIVIL GOVERNMENT were written by 38.___
A. Rene Descartes
B. John Locke
C. Thomas Hobbes
D. Jeremy Bentham
E. Benjamin Franklin

39. According to John Locke, a person acquires personal property by 39.___
A. taking it from others
B. asking God for use of it
C. needing it very much
D. mixing his labor with it
E. none of the above

40. According to Hobbes, life without a government would be 40.___
A. easygoing and fun
B. nasty, brutish and short
C. disorganized and chaotic
D. creative
E. all of the above

41. According to Hobbes, humans are by nature 41.___
I. selfish and greedy
II. arrogant and nasty
III. friendly and cooperative
IV. sociable and passive
V. healthy and happy

The CORRECT answer is:
A. I, II
B. II *only*
C. III *only*
D. III, IV
E. III, IV, V

42. According to Rousseau, humans are by nature 42.___
I. selfish and greedy
II. arrogant and nasty
III. friendly and cooperative
IV. sociable and passive
V. healthy and happy

The CORRECT answer is:
A. I *only*
B. I, II
C. II *only*
D. III, IV
E. III, IV, V

43. According to Freud, humans are by nature 43.___
I. selfish and greedy
II. arrogant and nasty
III. friendly and cooperative
IV. sociable and passive
V. healthy and happy

The CORRECT answer is:
A. I *only*
B. I, II
C. III *only*
D. III, IV, V
E. V *only*

44. The concept of the "general will" is *crucial* to the philosophy of 44.___
A. Hobbes
B. Hume
C. Rousseau
D. Locke
E. none of the above

45. A famous American thinker who *defended* American Independence from Great Britain was 45.___
I. Thomas Paine
II. Thomas Jefferson
III. Benjamin Franklin
IV. Herbert Spencer
V. all of the above

The CORRECT answer is:
A. I, II, III
B. I, III, IV
C. II, III
D. III, IV
E. V *only*

46. A famous British thinker who *refuted* Thomas Paine's ideas was 46.___
A. John Stuart Mill
B. Jeremy Bentham
C. Richard Price
D. Edmund Burke
E. none of the above

47. The American thinker responsible for popularizing the idea of civil disobedience was 47.___
A. Thomas Jefferson
B. George Washington
C. Benjamin Franklin
D. John Adams
E. Henry David Thoreau

48. The book WALDEN POND was written by 48.___
A. B. F. Skinner
B. Thomas Jefferson
C. John Quincy Adams
D. Henry David Thoreau
E. none of the above

49. In WALDEN POND, the MAIN subject of discussion was 49.___
A. fish
B. solitude
C. thinking and writing
D. religion
E. taxation

50. A civil disobedient *purposely* breaks the law in order to 50.___
A. cause trouble
B. reform an unjust law
C. change the system
D. create social unrest
E. all of the above

KEY (CORRECT ANSWERS)

1. C	11. E	21. C	31. E	41. A
2. B	12. B	22. A	32. E	42. E
3. A	13. C	23. B	33. C	43. B
4. E	14. D	24. D	34. D	44. C
5. C	15. A	25. E	35. B	45. C
6. D	16. D	26. C	36. C	46. D
7. D	17. E	27. B	37. C	47. E
8. D	18. C	28. E	38. B	48. D
9. E	19. C	29. D	39. D	49. E
10. C	20. B	30. E	40. B	50. B

EXAMINATION SECTION

TEST 1

DIRECTIONS: Each question or incomplete statement is followed by several suggested answers or completions. Select the one that BEST answers the question or completes the statement. *PRINT THE LETTER OF THE CORRECT ANSWER IN THE SPACE AT THE RIGHT.*

1. Utilitarianism is a very democratic moral theory because 1.___
   I. it does not rely upon judgments by experts
   II. it stipulates that everyone's utility be counted
   III. it stipulates that everyone determine his/her own utility
   IV. it rules out foolish or imprudent wishes and desires
   V. it rules out considerations of intentions

   The CORRECT answer is:
   A. I, II, IV
   B. I, II
   C. I, II, III
   D. II, III
   E. II, III, IV

2. The STRONG points of utilitarianism include 2.___
   I. its reliance on objective data
   II. its reliance on subjective data
   III. the fact that it applies to any and all situations
   IV. the fact that it prescribes a certain, 4-step procedure to determine what ought to be done, morally speaking
   V. its encouragement of social reforms

   The CORRECT answer is:
   A. I, III
   B. III, V
   C. I, III, V
   D. I, IV, V
   E. I, III, IV, V

3. The WEAK points of utilitarianism include 3.___
   I. its 4-step decision procedure
   II. its reliance on subjective data
   III. its reliance on objective data
   IV. that it applies to any and all situations
   V. its encouragement of social reforms

   The CORRECT answer is:
   A. I, III, IV
   B. I, II, IV
   C. I, II, V
   D. I, IV, V
   E. I, II, III

4. The utilitarianism decision procedure requires that each person 4.___
   I. figure out the consequences of each alternative action
   II. figure out which actions are available at the time
   III. figure out how much happiness each alternative will bring about
   IV. figure out how to maximize the happiness of everyone
   V. deliberate rationally over each act he/she performs

The CORRECT answer is:
A. I, III
B. I, II, III
C. I, III, V
D. I, II, III, V
E. all of the above

5. According to Kant, 5.___
I. lying is always wrong
II. lying is usually wrong, except in certain extreme situations
III. lying is the kind of action that harms people
IV. lying can lead to better consequences than not lying in some cases
V. lying cannot lead to better consequences than not lying in any case

The CORRECT answer is:
A. I, III
B. II, III
C. I, II
D. I, II, III
E. I, III, IV

6. According to John Stuart Mill, 6.___
I. lying is always wrong
II. lying is usually wrong, except in certain extreme situations
III. lying is the kind of action that harms people
IV. lying can lead to better consequences than not lying in some cases
V. lying cannot lead to better consequences than not lying in any case

The CORRECT answer is:
A. I, II
B. II, III
C. II, IV
D. I, II, IV
E. II, III, IV

7. Utilitarianism is opposed to paternalism because 7.___
I. it allows others to decide how happy a certain action will make an individual
II. it does not allow others to decide how happy a certain action will make an individual
III. it goes by majority rule
IV. it presupposes that utility is subjective
V. it presupposes that utility is objective

The CORRECT answer is:
A. I, II
B. II, III
C. II, IV
D. I, III, IV
E. II, III, IV

8. Deontology is opposed to paternalism because 8.___
I. it insists that individuals retain rights
II. it insists that each person should act according to his/her conscience
III. it insists that everyone ought to help others, even when others do not wish it
IV. autonomy requires free will in decision-making
V. all of the above

The CORRECT answer is:
A. I, II, III
B. I, II, V
C. I, II
D. I, II, IV
E. II, IV

9. Paternalism states that 9.___
A. everyone *always* ought to decide what is best for himself/herself as well as for others
B. it is sometimes right *not* to decide what is best for oneself and others
C. it is *never* right to decide what is best for oneself and others
D. it is *justifiable* to go against others' wishes if it is for their own good
E. it is *not* justifiable to go against others' wishes even if it is for their own good

10. Ethical absolutism is a type of theory that claims: 10.___
I. If an action is wrong at any one time, it is always wrong
II. Values vary in different cultures
III. It is morally wrong to make exceptions to moral rules
IV. It is morally justified to make exceptions to moral rules only if doing so is commanded by a legitimate authority
V. Relativism is incoherent and/or immoral

The CORRECT answer is:
A. I, III
B. III, V
C. I, V
D. I, III, V
E. I, III, IV, V

11. An absolutist ethical theory was held by 11.___
A. Jeremy Bentham
B. John Stuart Mill
C. Immanuel Kant
D. Charles Stevenson
E. all of the above

12. The Harm Principle states that 12.___
A. only determining the consequences of an act reveals whether that act is harmful or not
B. actions that harm are prima facie wrong
C. actions that harm are wrong
D. morality presupposes that everyone wants to avoid harming others
E. the only criterion of moral rightness is avoiding harm

13. Skepticism is a theory that states that 13.___
A. people frequently have ulterior motives for doing good actions
B. things usually turn out for the worst
C. knowledge is impossible
D. philisphical conclusions are only a matter of opinion
E. all of the above

14. Skeptical views were developed by 14.___
I. Plato
II. Aristotle
III. Descartes
IV. Hume
V. Kant

The CORRECT answer is:
A. III, IV
B. I, III
C. I, IV
D. III, IV, V
E. II, III, IV

15. Hume became famous for advocating that 15.___
I. miracles are not impossible
II. miracles are impossible
III. the principle of cause and effect is unfounded
IV. the principle of cause and effect is important and useful
V. that the existence of miracles does not justify faith in God

The CORRECT answer is:
A. I, III, V
B. II, III, IV
C. I, II, III
D. II, III, V
E. none of the above

16. Hume's epistemological theory may be *correctly* called 16.___
I. empiricist
II. idealist
III. materialist
IV. determinist
V. rationalist

The CORRECT answer is:
A. I, II, III
B. I, III, V
C. I, III
D. II, IV
E. I, III, IV

17. Kant's epistemological theory may be *correctly* called 17.___
I. empiricist
II. idealist
III. materialist
IV. determinist
V. rationalist

The CORRECT answer is:
A. I, III
B. I, V
C. I, II, III
D. II, III, V
E. II, V

18. Descartes' epistemological theory may be *correctly* called 18.___
I. empiricist
II. idealist
III. materialist
IV. determinist
V. rationalist

The CORRECT answer is:
A. I, II, III
B. I, III, V
C. I, II, V
D. II, IV
E. II, V

19. Rights are an important concept in one of the following moral theories: 19.___
   A. Utilitarianism
   B. Deontology
   C. Emotivism
   D. Relativism
   E. Egoism

20. Egoism is a moral theory that states that 20.___
   A. everyone always acts so as to maximize his/her own self-interest
   B. everyone always ought to act so as to maximize his/her own self-interest
   C. doing what is good for others requires self-sacrifice
   D. altruistic acts are few and far between
   E. All of the above

21. Egoistic ethical theory includes the following claims: 21.___
   I. Self-interest is the only worthwhile motive for doing an act
   II. Altruistic acts are always done to promote self-interest
   III. Human beings are rational in that they seek to maximize their own self-interest
   IV. God helps those who help themselves
   V. To be morally good, all we need to do is be prudent

   The CORRECT answer is:
   A. I, II, III
   B. I, II, III, IV
   C. I, II, III, V
   D. II, III, V
   E. All of the above

22. Egoism as a moral theory was advocated by 22.___
   A. Aristotle
   B. Plato
   C. Kant
   D. Hobbes
   E. Hume

23. Psychological egoism states that 23.___
   I. everyone always ought to promote his/her own self-interest
   II. no one should try to promote another person's self-interest
   III. everyone always tries to promote his/her own self-interest
   IV. everyone usually tries to promote his/her own self-interest
   V. no one ever tries to promote another person's self-interest

   The CORRECT answer is:
   A. II, III
   B. III, V
   C. III, IV, V
   D. II, III, V
   E. I, III, V

24. Some fields that assume psychological egoism are 24.___
   I. psychology
   II. political science
   III. economics
   IV. literature
   V. religion

The CORRECT answer is:
A. I, II
B. II, III
C. II, III, IV
D. II, III, V
E. II, IV

25. Philosophers who have emphasized the importance of habit in human behavior is(are) 25.___
I. Plato
II. Aristotle
III. Hume
IV. Hobbes
V. Descartes

The CORRECT answer is:
A. I *only*
B. II *only*
C. I, II
D. I, III
E. II, III

26. Epicurus is famed for his theory concerning 26.___
I. good food
II. Epicurean delights other than good food
III. hedonism
IV. living a life that brings peace of mind
V. all of the above

The CORRECT answer is:
A. I, III
B. I, IV
C. III, IV
D. II, III
E. II, III, IV

27. Epicurus' version of moral theory recommends 27.___
I. not getting angry at frustrating situations
II. avoiding scholarly work because it leads to unhappiness
III. studying science as much as possible to understand reality
IV. living on bread and water to become healthy
V. eating good food for pleasure and happiness

The CORRECT answer is:
A. I, II
B. I, II, V
C. III, IV
D. I, II, III
E. I, II, IV

28. An "analytical" statement is *one* that 28.___
A. requires deep philosophical reflection
B. requires the use of mathematics
C. provides information about the world
D. defines a term
E. cannot be false

29. A "synthetic" statement is *one* that 29.___
A. is invented by a person
B. requires the use of mathematics
C. provides information about the world
D. defines a term
E. cannot be false

30. A "synthetic a priori" statement is one that is 30.___
A. informative and provides a definition
B. *not* informative and provides a definition
C. informative and provides no definition
D. informative and cannot be false
E. informative and may or may not be false

31. According to David Hume, our notion that similar causes bring about similar effects is really reducible to 31.___
I. wishful thinking
II. practical experience
III. custom and habit
IV. ancient philosophical ideas
V. all of the above

The CORRECT answer is:
A. II, III
B. I, III
C. III, IV
D. I, II, III
E. I, II, III, IV

32. The allegory of the cave, in Plato's REPUBLIC, concerns 32.___
I. slaves rebelling against their masters
II. epistemological problems of knowledge and illusion
III. psychological habits and beliefs
IV. the danger of thinking philosophically
V. people's natural impulse to learn as much as possible

The CORRECT answer is:
A. I, II, III
B. II, III, V
C. I, III, V
D. II, III
E. II, III, IV

33. In his book ON LIBERTY, John Stuart Mill argued that 33.___
I. the majority should make moral and political decisions, not an elite group
II. individuals have rights against other individuals and minorities
III. individuals ought to go along with the majority
IV. in democratic societies, governments ought not to meddle with individuals who are not harming others
V. individuals are subject to the tyranny of the majority

The CORRECT answer is:
A. IV, V
B. I, II
C. I, IV
D. III, IV
E. III, IV, V

34. John Stuart Mill attributes most of the ideas in ON LIBERTY to 34.___
A. Jeremy Bentham
B. James Mill (his father)
C. Harriet Taylor
D. David Hume
E. John Locke

35. Plato thought that virtue was 35.___
I. the most important thing in life
II. difficult to define
III. something you should teach to children from an early age
IV. something that cannot be taught
V. achieved if and when justice was achieved

The CORRECT answer is:
A. II, IV
B. I, II
C. I, II, III
D. I, II, IV
E. I, IV

36. The "use/mention" distinction is important because 36.___
I. it permits the analysis of meaning
II. we sometimes talk about a word
III. it requires using inverted commas
IV. it helps philosophical analysis
V. otherwise we will commit equivocations

The CORRECT answer is:
A. I, II
B. II, IV
C. I, III, V
D. I, II, IV, V
E. all of the above

37. "Meta-ethics" concerns 37.___
A. moral judgments about what is right and what is wrong
B. empirical judgments about moral beliefs
C. philosophical analysis of moral terms
D. logical critiques of philosophical arguments in ethics
E. all of the above

38. Stipulative definitions are definitions that 38.___
I. specify the meaning of a term according to the author
II. use standard specifications of meaning such as a dictionary
III. rely upon standard usage of a language
IV. create uniformity and standard usage
V. are designed to clarify complex problems

The CORRECT answer is:
A. I, II
B. IV, V
C. I, III, V
D. I, IV, V
E. I, V

39. Theology involves 39.___
I. epistemological distinctions
II. metaphysical claims
III. study of religious dogma
IV. philosophy of religion
V. theoretical justifications of religious beliefs

The CORRECT answer is:
A. I, II, III
B. II, III, IV
C. I, III, V
D. I, III, IV
E. III, IV, V

40. A criticism of utilitarianism as a moral theory is 40.___
A. that it requires a knowledge of the future to make a decision
B. that it makes the same action right at one time and wrong at another
C. that it is unjust
D. that it is very time-consuming
E. all of the above

41. The problem with the following statement, "It is wrong to make moral judgments" is that 41.___
    A. we do so all the time   B. it is arrogant
    C. it is a contradiction   D. it leads to apathy
    E. All of the above

42. According to Aristotle, an enthymeme is 42.___
    A. a form of rhetorical argument
    B. a common method of persuasion
    C. ruled out in legal arguments
    D. a syllogism missing a premise
    E. All of the above

43. Playing "Devil's Advocate" requires that a person 43.___
    A. study religious dogma to know what the Devil believes
    B. retain a firm hold on his/her beliefs
    C. argue against his/her own opinion for the sake of the argument
    D. follow Socrates' example and ask questions all the time
    E. All of the above

44. The art of forensics involves 44.___
    I. chemical analysis of criminal evidence
    II. arguments for and against someone accused of something
    III. debating
    IV. logical prowess
    V. All of the above

    The CORRECT answer is:
    A. II, III, IV   B. I, II, III
    C. III, IV, V   D. II, III
    E. III, IV

45. Included in Aristotle's list of the ingredients for happiness is(are): 45.___
    I. High social station   II. Friends, children
    III. Athletic ability   IV. Beautiful looks
    V. Money

    The CORRECT answer is:
    A. I *only*   B. II *only*
    C. III *only*   D. IV *only*
    E. all of the above

46. An oligarchy is a form of government in which 46.___
    A. only the elite rule   B. only aristocrats rule
    B. kings and nobility rule   D. property owners rule
    E. no one rules

47. According to Marx and Engels, 47.___
    I. capitalism was a liberating influence in its early days
    II. society is the history of class struggles
    III. dominant classes inevitably win out in political struggles
    IV. labor is intrinsically immoral under capitalism
    V. all of the above

The CORRECT answer is:
A. I, III, V
B. I, II
C. II, IV
D. I, II, IV
E. I, II, III

48. John Locke provided a justification of 48.___
I. the modern state
II. private property
III. separating God from the government
IV. owning apples one has picked oneself
V. all of the above

49. Among the "natural rights" emphasized by John Locke is (are): 49.___
I. The right to keep what one has labored for
II. The right to pursue happiness
III. The right to have one's property protected by government
IV. The right to control our own persons
V. All of the above

The CORRECT answer is:
A. I, II, III
B. II, III, IV
C. I, III, IV
D. I, II, III, IV
E. V *only*

50. *One* of the MAIN concepts in existentialism, as promoted by Jean-Paul Sartre, is 50.___
A. pessimism
B. futility
C. suicide
D. choice
E. faith

KEY (CORRECT ANSWERS)

1. C	11. C	21. C	31. A	41. E
2. E	12. B	22. D	32. E	42. D
3. B	13. C	23. B	33. A	43. C
4. E	14. A	24. B	34. C	44. A
5. A	15. D	25. E	35. D	45. E
6. C	16. E	26. C	36. D	46. D
7. C	17. E	27. C	37. C	47. D
8. D	18. E	28. D	38. E	48. D
9. D	19. B	29. C	39. E	49. C
10. D	20. B	30. D	40. E	50. D

TEST 2

DIRECTIONS: Each question or incomplete statement is followed by several suggested answers or completions. Select the one that BEST answers the question or completes the statement. *PRINT THE LETTER OF THE CORRECT ANSWER IN THE SPACE AT THE RIGHT.*

1. The view that each thing or person tends toward a goal or purpose is 1.___
   I. emotivism
   II. functionalism
   III. teleology
   IV. idealism
   V. materialism

   The CORRECT answer is:
   A. I *only*
   B. II *only*
   C. III, IV
   D. II, III
   E. IV, V

2. The philosopher FIRST responsible for a teleological view of life is 2.___
   I. Plato
   II. Socrates
   III. Aristotle
   IV. Heraclitus
   V. Democritus

   The CORRECT answer is:
   A. I *only*
   B. I, II
   C. III *only*
   D. IV *only*
   E. V *only*

3. The philosopher frequently called a "peripatetic" and whose Academy was also referred to as "The Peripatetic" school was 3.___
   I. Plato
   II. Socrates
   III. Aristotle
   IV. Heraclitus
   V. Plotinus

   The CORRECT answer is:
   A. I *only*
   B. II *only*
   C. I, II
   D. III *only*
   E. IV, V

4. The view that society is composed of an aggregate of units is called 4.___
   I. functionalism
   II. teleology
   III. collectivism
   IV. organicism
   V. individualism

   The CORRECT answer is:
   A. I *only*
   B. I, II
   C. III *only*
   D. IV *only*
   E. V *only*

5. The view that society is comprised of integrally related parts and that the whole is greater than the sum of its parts is called 5.___

I. functionalism
II. teleology
III. collectivism
IV. organicism
V. individualism

The CORRECT answer is:

A. I, II
B. III *only*
C. III, IV
D. IV *only*
E. V *only*

6. Philosophers who were organicists were 6.___

I. Democritus
II. Aristotle
III. Bentham
IV. Hegel
V. Marx

The CORRECT answer is:

A. I *only*
B. I, II
C. III *only*
D. II, IV, V
E. IV *only*

7. Philosophers who were individualists were 7.___

I. Heraclitus
II. Aristotle
III. Bentham
IV. Hegel
V. Marx

The CORRECT answer is:

A. I *only*
B. I, II
C. III *only*
D. IV, V
E. V *only*

8. Philosophers who agreed that happiness was the *primary* goal in human life are 8.___

I. Socrates
II. Aristotle
III. Descartes
IV. Bentham
V. John Stuart Mill

The CORRECT answer is:

A. I *only*
B. I, III
C. II *only*
D. II, IV, V
E. IV, V

9. Plato and Aristotle *primarily* DISAGREED about the philosophical issue of 9.___

I. empirical evidence
II. cosmology
III. politics
IV. whether virtue was attainable
V. aesthetics

The CORRECT answer is:

A. I *only*
B. II *only*
C. I, II, V
D. II, IV
E. III *only*

10. Aristotle emphasized, in direct contrast to Plato, that 10.___
I. most people should train in philosophy
II. people ought to actively pursue goals
III. studying virtue is the best thing to do
IV. individual acts have great moral significance
V. universal goodness is the source of each good action

The CORRECT answer is:
A. I *only*
B. II, III
C. IV *only*
D. II, III, IV
E. II, V

11. The philosopher in ancient Greece who thought that a person who is ugly, poor, and without children has little chance of being happy is 11.___
I. Socrates
II. Plato
III. Aristotle
IV. Pythagoras
V. Plotinus

The CORRECT answer is:
A. I *only*
B. I, II
C. III *only*
D. III, IV
E. IV, V

12. Philosophers who thought that studying morality is worthwhile only if it makes the student more ethical are 12.___
I. Plato
II. Aristotle
III. Hegel
IV. Descartes
V. Marx

The CORRECT answer is:
A. I *only*
B. I, III
C. II, IV
D. II, V
E. III, V

13. The philosopher MOST responsible for the idea that the good life is that which emphasizes moderation, i.e., "the mean between the extremes", is 13.___
I. Plato
II. Aristotle
III. Hegel
IV. Leibniz
V. Spinoza

The CORRECT answer is:
A. I *only*
B. II *only*
C. I, III
D. III, IV
E. V *only*

14. "Pyrrhonism" refers to a philosophy emphasizing 14.___
I. love
II. justice
III. asceticism
IV. heroic skepticism
V. hedonism

The CORRECT answer is:
A. I *only*
B. I, II
C. III *only*
D. III, IV
E. IV, V

15. Pythagoreans emphasized in their philosophy that 15.___
I. happiness is the goal of life
II. the pure life was the best
III. philosophical discussion is preferable to silence
IV. silence is preferable to philosophical discussion
V. asceticism is desirable

The CORRECT answer is:
A. I *only*
B. I, V
C. II *only*
D. III, V
E. II, IV, V

16. Axiology is that realm of philosophy which concerns 16.___
I. metaphysics
II. reality
III. love and justice
IV. goodness
V. pleasure

The CORRECT answer is:
A. I *only*
B. I, II
C. III *only*
D. IV *only*
E. V *only*

17. *Basic* features of philosophy include 17.___
I. generality
II. logic
III. emotion
IV. particularity
V. interpretation

The CORRECT answer is:
A. I, III
B. II *only*
C. I, II, IV
D. I, II, V
E. III, IV

18. A noteworthy medieval philosopher was 18.___
A. Plato
B. Hobbes
C. Augustine
D. Plotinus
E. Thucydides

19. Modern philosophers, as compared to ancient and medieval philosophers, emphasize 19.___
I. theology
II. metaphysical speculation
III. logical analysis
IV. linguistic analysis
V. ordinary language

The CORRECT answer is:
A. I *only*
B. I, II
C. III *only*
D. III, IV
E. III, IV, V

20. Studying philosophy contributes to one's life in that it 20.___
I. fosters intellectual growth
II. reaffirms one's early beliefs
III. is culturally enriching
IV. guarantees that one will be wary of falsehoods
V. satisfies our curiosity

The CORRECT answer is:
A. I, III, V
B. II, IV
C. I, III, IV
D. II, III, IV
E. II, V

21. Some characteristics of the deductive method in philosophy are that it 21.___
I. is intuitive
II. generates highly probable conclusions
III. generates certain conclusions
IV. uses axioms
V. relies upon empirical data

The CORRECT answer is:
A. I, III, IV
B. II, V
C. III, IV
D. III, V
E. II, III, IV, V

22. The primary difference between ontology and cosmology is that 22.___
I. the former is logically independent of the latter
II. one is concerned with being and the other with conclusions
III. the latter is concerned with origins and the former with what is
IV. only one is fundamentally connected to metaphysics
V. only one concerns the origins of the universe

The CORRECT answer is:
A. I *only*
B. II, IV
C. I, III, V
D. III, V
E. IV, V

23. One MAJOR difference between materialism and idealism is that 23.___
I. only one represents a major philosophical school
II. the latter includes a non-physical element
III. the former includes no supersensible rationale
IV. the ancients only studied the latter
V. only one is in accordance with common sense

The CORRECT answer is:
A. I, II
B. II *only*
C. I, IV
D. II, III
E. III, IV, V

24. The FIRST philosopher to claim that planets moved in an elliptical, not circular, manner was 24.___
I. Democritus
II. Ptolemy
III. Galileo
IV. Copernicus
V. Descartes

The CORRECT answer is:
A. I *only*
B. II *only*
C. III, IV
D. III *only*
E. IV, V

25. The view that all of reality can be reduced to one substance or entity is 25.___

I. monism
II. dualism
III. idealism
IV. animism
V. pantheism

The CORRECT answer is:
A. I *only*
B. I, III
C. II *only*
D. III *only*
E. IV, V

26. Philosophers who are "pluralists" are those who claim that 26.___

I. no one knows what is true or false
II. seeing is believing
III. material reality is reducible to more than one element
IV. material reality is reducible to one element
V. there is no such thing as material reality

The CORRECT answer is:
A. I *only*
B. I, III
C. III *only*
D. IV *only*
E. V *only*

27. Frequently, monists *also* claim that 27.___

I. only minds or spirits are real
II. philosophy should be abandoned for science
III. material reality is an illusion
IV. appearance is more important than reality
V. reality is more important than appearance

The CORRECT answer is:
A. I *only*
B. I, II
C. III *only*
D. I, III, IV
E. none of the above

28. One claim *frequently* put forth by "animists" is that 28.___

I. only animals embody religious spirits
II. living plants and animals each represent God directly
III. living plants and animals have a special spirit
IV. living plants and animals are at man's disposal
V. humans are the only creatures animated by God's spirit

The CORRECT answer is:
A. I *only*
B. I, IV
C. II, III
D. III, V
E. IV *only*

29. The fundamental meaning of "hylozoism" is represented by the view that 29.___

I. water is the creative force of the universe
II. life is inseparable from matter
III. life must be separated from inanimate matter
IV. atomic particulars are the nucleus of matter
V. there is no such thing as inanimate matter

The CORRECT answer is:
A. I *only*
B. I, II
C. II, V
D. III, IV
E. IV, V

30. The Latin phrase "sui generis" *usually* refers to 30.___
I. self-generation
II. soft cells in atomic matter
III. the generation of animal life
IV. God's existence
V. human existence

The CORRECT answer is:
A. I *only*
B. II *only*
C. I, III
D. I, IV
E. I, III, V

31. The cosmological theory called "archebiosis" refers to the view that 31.___
I. most natural conditions are created by God's will
II. nature is largely the result of a chance combination of events
III. nature has a rational explanation
IV. natural life is explainable by teleological purposes
V. life's origins must remain mysterious to humans

The CORRECT answer is:
A. I *only*
B. II *only*
C. I, IV
D. III *only*
E. V *only*

32. Darwin's fundamental principle of scientific explanation for biological life is 32.___
I. the principle of natural selection
II. the principle of immutable force
III. designed to explain extinct species
IV. rarely verified through empirical evidence
V. equally applicable to human and legal society

The CORRECT answer is:
A. I *only*
B. II, III
C. I, III
D. II, III, IV
E. IV, V

33. The theological view that challenges Darwin's theory is 33.___
I. primordialism
II. fundamentalism
III. episcopalianism
IV. asceticism
V. creationism

The CORRECT answer is:
A. I *only*
B. I, III
C. II *only*
D. IV, V
E. II, V

34. According to David Hume, the human mind is 34.___
A. the organizing principle of human life
B. a collection of perceptions
C. indubitably given a priori
D. ultimately indistinguishable from matter
E. an immaterial substance

35. According to Aristotle, the human mind is 35.___
A. the organizing principle of human life
B. a collection of perceptions
C. indubitably given a priori
D. ultimately indistinguishable from matter
E. an immaterial substance

36. According to Descartes and Berkeley, the human mind is 36.___
I. the organizing principle of human life
II. a collection of perceptions
III. indubitably given a priori
IV. ultimately indistinguishable from matter
V. an immaterial substance

The CORRECT answer is:
A. I *only*
B. I, II
C. III, V
D. IV *only*
E. I, III, IV

37. If dualism is TRUE, then 37.___
I. none of us know if we have a body
II. none of us know if we have a mind
III. the relationship of mind to body remains a question
IV. Descartes was correct
V. Marx was correct

The CORRECT answer is:
A. I *only*
B. I, II
C. III *only*
D. III, IV
E. V *only*

38. The theory called "interactionism" claims that 38.___
I. Hegel's theory of history is correct
II. the mind is separate from the body
III. the mind and the body are subtly inter-related
IV. the mind and the body are identical
V. the mind actually is no more than the brain

The CORRECT answer is:
A. I *only*
B. II *only*
C. III *only*
D. I, III
E. IV, V

39. The theory of mind called "parallelism" claims that 39.___
I. God's will is parallel to man's
II. minds and bodies never interact
III. no one knows for sure if the self is intact
IV. theories of theology amount to the same thing
V. minds are no more than brains

The CORRECT answer is:
A. I *only*
B. II *only*
C. I, III
D. IV *only*
E. V *only*

40. The theory of mind called "identity theory" claims that 40.___
I. God's will is identical to man's
II. no one knows whether the mind is real or not
III. theories of theology amount to the same thing
IV. minds and bodies never interact
V. minds are no more than brains

The CORRECT answer is:
A. I *only*
B. II *only*
C. III *only*
D. III, IV
E. V *only*

41. The peculiarity of the theory called "epiphenomenalism" in the philosophy of mind is its claim that 41.___
I. the mind is no more than the brain
II. the body has no independent reality from the mind
III. the mind has no independent reality from the body
IV. there are no bodies, only minds
V. theology is the source of identity

The CORRECT answer is:
A. I *only*
B. II *only*
C. III *only*
D. II, IV
E. V *only*

42. The theory called "fatalism" claims that 42.___
I. the Fates are in contact with human souls
II. whatever happens is meant to happen
III. actions determine to a large extent a person's life
IV. events are largely uncontrolled by human action
V. people deserve what they get

The CORRECT answer is:
A. I *only*
B. I, V
C. II *only*
D. II, IV
E. III, IV, V

43. The theory called "predestination" claims that 43.___
I. the Fates are in contact with human souls
II. events that occur are planned by an extra-human force
III. actions are largely determined by human actions
IV. what you do has no effect on what happens
V. there is a force larger than human powers in control of things

The CORRECT answer is:
A. I, II
B. II, III
C. II, V
D. II, IV, V
E. IV, V

44. *Determinism* is the philosophical view that 44.___
I. no one believes could possibly be true
II. determined people can accomplish anything
III. human actions are explainable by preceding causes
IV. human actions are not explainable by preceding causes
V. serious motivation solves most human problems

The CORRECT answer is:
A. I *only*
B. I, V
C. II *only*
D. III, V
E. III *only*

45. The theory called "indeterminism" claims that 45.___
I. determined people are ineffectual against the fates
II. human actions are explainable by preceding causes
III. human actions are not explainable by preceding causes
IV. humans have free will
V. God controls human events to a large degree

The CORRECT answer is:
A. I *only*
B. I, III
C. III, IV
D. II, V
E. V *only*

46. The school(s) of thought famous for espousing indeterminism is(are) 46.___
I. existentialism
II. empiricism
III. natural law theory
IV. teleological theory
V. identity theory

The CORRECT answer is:
A. I *only*
B. I, II
C. I, III
D. IV *only*
E. IV, V

47. The difference between *objective* and *subjective* idealists is that the 47.___
I. former believe in reality, whereas the latter do not
II. former believe that ideas are eternal
III. latter believe that reality is independent of the mind
IV. latter believe that reality is whatever is perceived by minds
V. former believe that reality is whatever is perceived by minds

The CORRECT answer is:
A. I *only*
B. II, IV
C. I, III
D. V *only*
E. II, III, IV

48. One problem with philosophies based on common sense is that 48.___
I. sticks do not bend when placed in a glass of water
II. science defies many principles of common sense
III. so few people have common sense
IV. philosophical theories are supposed to be complex
V. they are so plentiful

The CORRECT answer is:
A. I *only*
B. I, III
C. I, II
D. III, IV
E. IV, V

49. The philosophical school of pragmatism asserts that 49.___
I. learning takes place within a physical environment
II. people have innate knowledge
III. appearances are never the same as realities
IV. experience is fundamental
V. logic is paramount in learning

The CORRECT answer is:
A. I *only*
B. I, III
C. II, III
D. I, IV
E. V *only*

50. John Dewey's form of pragmatism claimed that 50.___
I. knowledge reflects the world
II. human knowledge reshapes the world
III. knowledge is an instrumental thing
IV. knowledge is intrinsically good
V. science ought to be rejected in favor of pragmatism

The CORRECT answer is:
A. I *only*
B. II *only*
C. I, IV
D. II, III
E. V *only*

KEY (CORRECT ANSWERS)

1. D	11. C	21. C	31. B	41. C
2. C	12. D	22. D	32. C	42. D
3. D	13. B	23. D	33. E	43. D
4. E	14. D	24. B	34. B	44. E
5. C	15. E	25. A	35. A	45. C
6. D	16. D	26. C	36. C	46. C
7. C	17. D	27. E	37. D	47. B
8. D	18. C	28. C	38. C	48. C
9. A	19. E	29. C	39. B	49. D
10. D	20. A	30. D	40. E	50. D

EXAMINATION SECTION

TEST 1

DIRECTIONS: Each question or incomplete statement is followed by several suggested answers or completions. Select the one that BEST answers the question or completes the statement. *PRINT THE LETTER OF THE CORRECT ANSWER IN THE SPACE AT THE RIGHT.*

1. American philosophers include 1.___
   I. Bertrand Russell
   II. John Stuart Mill
   III. William James
   IV. John Dewey
   V. Alfred North Whitehead

   The CORRECT answer is:
   A. I *only*
   B. I, II, V
   C. II, IV, V
   D. III, IV, V
   E. II, IV

2. Correspondence theory is *primarily* used as a 2.___
   I. mathematical assumption
   II. axiomatic presupposition
   III. test of the truth of statements
   IV. limitation on human knowledge
   V. theory of art

   The CORRECT answer is:
   A. I *only*
   B. I, II
   C. III *only*
   D. II, IV
   E. III, V

3. Phenomenalism is a theory that claims that 3.___
   I. reality is autonomous and independent of human minds
   II. reality is the joint product of minds and sensory data
   III. some philosophers will know what noumena are
   IV. no one ever knows what noumena are
   V. phenomena and noumena are distinguishable

   The CORRECT answer is:
   A. I *only*
   B. II *only*
   C. I, III
   D. II, V
   E. II, IV

4. The philosopher *primarily* responsible for distinguishing noumena and phenomena is 4.___
   A. Hume
   B. Aristotle
   C. Kant
   D. Descartes
   E. Dewey

5. The philosophical theory called "coherence theory" is one that 5.___
   I. only materialists believe
   II. only idealists believe
   III. relies for its basis on other theories
   IV. emphasizes soundness over validity
   V. emphasizes validity over soundness

The CORRECT answer is:
A. II, III
B. I, IV
C. I, IV, V
D. II, III, V
E. III, V

6. Absolutist theory is *primarily* found in theories concerning 6.___
A. epistemology
B. metaphysics
C. ethics
D. aesthetics
E. epiphenomenalism

7. The "Theory of the Mean" was created by 7.___
I. Plato
II. Aristotle
III. Descartes
IV. Kant
V. Mill

The CORRECT answer is:
A. I *only*
B. II *only*
C. II, III
D. IV *only*
E. V *only*

8. The term "mean" in the moral "Theory of the Mean" refers to a(n) 8.___
I. quantitative relationship
II. qualitative relationship
III. absolute figure
IV. relative figure
V. low or vulgar person or thing

The CORRECT answer is:
A. I *only*
B. I, III
C. I, IV
D. II, V
E. V *only*

9. The term "theologism" refers *primarily* to ____ religious theory. 9.___
I. Hindu
II. Buddhist
III. Christian
IV. Taoist
V. Confucianist

The CORRECT answer is:
A. I *only*
B. I, III
C. II *only*
D. III *only*
E. I, IV, V

10. The term "formalism" applies BEST to the theory of 10.___
I. Aristotle
II. Hume
III. Kant
IV. Mill
V. Dewey

The CORRECT answer is:
A. I *only*
B. II, IV
C. III *only*
D. IV, V
E. V *only*

11. Motives are MOST important in the moral theory developed by 11.___

I. Aristotle
II. Plato
III. Descartes
IV. Kant
V. Mill

The CORRECT answer is:
A. I *only*
B. II, IV
C. IV *only*
D. III, V
E. IV, V

12. The term "eudaemonism" in moral theory refers to 12.___

I. pain
II. pleasure
III. happiness
IV. good will
V. hedonism

The CORRECT answer is:
A. I *only*
B. I, V
C. II, III
D. III *only*
E. IV *only*

13. The cognitivist in moral theory claims that 13.___

I. no one knows what is right and wrong
II. moral judgments are subjective
III. moral judgments are largely unprovable
IV. moral judgments are *either* true or false
V. moral judgments are *neither* true or false

The CORRECT answer is:
A. I *only*
B. I, II
C. III, V
D. IV *only*
E. V *only*

14. The *primary* concept(s) in the philosophy of aesthetics is(are) 14.___

I. virtue
II. beauty
III. goodness
IV. right
V. justice

The CORRECT answer is:
A. I *only*
B. I, II
C. III, IV
D. II *only*
E. IV, V

15. For Plato, the MAJOR problem with art was that it 15.___

I. moved people to drastic action
II. inspired moral righteousness
III. was too difficult for all but a few to master
IV. is too trivial for political philosophy
V. wasted philosophers' time

The CORRECT answer is:
A. V *only*
B. I *only*
C. II, III
D. III, IV
E. I, III, V

16. The philosopher(s) who claimed that art was an imitation of nature was(were) 16.____
I. Kant
II. Aristotle
III. Dewey
IV. Augustine
V. Santayana

The CORRECT answer is:
A. I *only*
B. I, III
C. II *only*
D. I, IV, V
E. IV, V

17. Throughout the Middle Ages, art was 17.____
I. held in philosophical disrespect
II. irrelevant to religion
III. designed to portray religious sentiments
IV. banned entirely
V. used to encourage moral behavior

The CORRECT answer is:
A. I *only*
B. I, II
C. III *only*
D. IV *only*
E. III, V

18. The philosopher(s) who claimed that the purpose of art is to give pleasure was(were) 18.____
I. Plato
II. Aristotle
III. Santayana
IV. Kant
V. Dewey

The CORRECT answer is:
A. I *only*
B. I, IV
C. III *only*
D. IV *only*
E. II, V

19. The philosopher(s) who claimed that art concerns the character of humans was(were) 19.____
I. Plato
II. Aristotle
III. Santayana
IV. Kant
V. Dewey

The CORRECT answer is:
A. I *only*
B. I, III
C. II *only*
D. I, IV, V
E. V *only*

20. The philosopher(s) who thought that art should have social significance was(were) 20.____
I. Plato
II. Aristotle
III. Santayana
IV. Kant
V. Dewey

The CORRECT answer is:
A. I *only*
B. II, V
C. III *only*
D. V *only*
E. IV, V

21. The philosopher(s) who thought that the most beautiful art symbolized moral goodness was(were) 21.___
I. Plato
II. Aristotle
III. Santayana
IV. Kant
V. Dewey

The CORRECT answer is:
A. I *only*
B. II, V
C. III *only*
D. IV *only*
E. I, III, V

22. According to the philosopher Schopenhauer, art is *primarily* 22.___
I. a means to a moral end
II. aimed at social significance
III. used as an escape
IV. designed to make people better citizens
V. a reflection of nature

The CORRECT answer is:
A. I *only*
B. I, IV
C. II *only*
D. III *only*
E. I, III, V

23. Benedetto Croce, the aesthetician, believed that art is 23.___
I. therapeutic
II. an enrichment of everyday life
III. found in normal, everyday experiences
IV. an intuition of ultimate reality
V. spiritual

The CORRECT answer is:
A. I *only*
B. I, III
C. II, III
D. IV *only*
E. IV, V

24. The only philosopher to suggest that art be banned in society is 24.___
A. Plato
B. Aristotle
C. Dewey
D. Santayana
E. Croce

25. Among other philosophers and thinkers, the one(s) who believe that religious devotion is based *primarily* upon human reason is(are) 25.___
I. Luther
II. Calvin
III. Kant
IV. Nietzsche
V. Kierkegaard

The CORRECT answer is:
A. I *only*
B. II *only*
C. I, II
D. III *only*
E. IV, V

26. Cosmological arguments for the existence of God are *primarily* based on 26.___
I. faith
II. reason
III. nature
IV. God's will
V. a priori logic

The CORRECT answer is:
A. I *only*
B. I, IV
C. II, V
D. III *only*
E. I, III, V

27. Ontological arguments for the existence of God are *primarily* based on 27.___
I. faith
II. reason
III. nature
IV. God's will
V. a priori logic

The CORRECT answer is:
A. I *only*
B. I, IV
C. II, V
D. III *only*
E. I, III, V

28. Teleological arguments for the existence of God are *primarily* based on 28.___
I. divine revelation
II. faith
III. reason
IV. the design of nature
V. a priori logic

The CORRECT answer is:
A. I *only*
B. I, II
C. III, V
D. IV *only*
E. I, II, V

29. The claim that God is *omniscient* means that He is 29.___
A. without a beginning or an end
B. all-powerful
C. everywhere at all times
D. all-knowing
E. the source of grace

30. The claim that God is *omnipresent* means that He is 30.___
I. without a beginning or an end
II. all-powerful
III. everywhere at all times
IV. all-knowing
V. the source of grace

The CORRECT answer is:
A. I *only*
B. II *only*
C. I, IV
D. III *only*
E. V *only*

31. The claim that God is *omnipotent* means that He is 31.___
I. without a beginning or an end
II. all-powerful
III. everywhere at all times
IV. all-knowing
V. the source of grace

The CORRECT answer is:
A. I *only*
B. I, III
C. II *only*
D. IV *only*
E. IV, V

32. The claim that no one can deserve to go to Heaven indicates that God is 32.___
I. without a beginning or an end
II. all-powerful
III. everywhere at all times
IV. all-knowing
V. the source of grace

The CORRECT answer is:
A. I *only*
B. I, III
C. II, IV
D. IV *only*
E. V *only*

33. One problem concerning God's infinite goodness is that 33.___
I. no one has ever seen God
II. Jesus spoke in vague parables
III. there is so much evil in the world
IV. people have no free will
V. disasters like the Holocaust increase religious skepticism

The CORRECT answer is:
A. I *only*
B. I, II
C. III *only*
D. III, V
E. I, III, V

34. The thinker and philosopher Siddhartha Gautama founded the religion of 34.___
A. Hinduism
B. Buddhism
C. Taoism
D. Confucianism
E. Theism

35. The Buddha was a thinker, teacher and philosopher who 35.___
I. was born poor and humble
II. was born rich and powerful
III. advocated rejecting desires
IV. denied the existence of the self
V. claimed that God resides in the self

The CORRECT answer is:
A. I *only*
B. I, V
C. II, V
D. II, III
E. II, III, IV

36. Buddha claimed that the ultimate goal of every human being is to 36.___
I. live the most pleasurable life
II. attain Nirvana
III. reject all pleasures of life
IV. use moderation in all activities
V. worship God

The CORRECT answer is:
A. I *only*
B. I, V
C. II, III
D. II, IV
E. II, V

37. A *primary* concept in Judaism is 37.___
I. the grace of God
II. the covenant
III. God's infinite love
IV. the law of God
V. revelation

The CORRECT answer is:
A. I, V
B. II *only*
C. II, IV
D. II, III
E. III, V

38. Fundamental concepts in Catholicism are 38.___
I. the law of God
II. God's divinity in Christ
III. the Trinity
IV. revelation
V. sacraments

The CORRECT answer is:
A. I *only*
B. I, III
C. II, III, IV
D. II, III, V
E. I, V

39. The religion of Islam is based on 39.___
I. the Koran
II. an afterlife of spiritual existence
III. the prophet Mohammed
IV. God's infinite mercy
V. a Heaven full of sensual pleasures

The CORRECT answer is:
A. I *only*
B. I, II
C. I, III
D. II, IV
E. I, III, V

40. Protestantism focuses *primarily* upon 40.___
I. individual dignity
II. a personal God
III. sacraments
IV. intermediaries between man and God
V. attaining moral behavior based on Christ's teachings

The CORRECT answer is:
A. I *only*
B. II, III
C. I, II, IV
D. I, II, V
E. I, III, IV

41. The *primary* problem with the claim "an act such as murder is wrong because God says it is," is that 41.___
I. some people think murder is right
II. it is difficult to know what God says
III. morality can be observed independently of God's word
IV. atheists cannot be moral
V. religious people must use extreme self-discipline to avoid sin

The CORRECT answer is:
A. I, III
B. II *only*
C. II, IV
D. III, V
E. III, IV, V

42. The problem with the claim, "God says that murder is wrong because it is," is that 42.___
I. some people think murder is right
II. it is difficult to know what God says
III. morality can be observed independently of God's word
IV. atheists cannot be moral
V. religious people must use extreme self-discipline to avoid sin

The CORRECT answer is:
A. I *only*
B. I, II
C. III *only*
D. II, IV
E. I, III, V

43. Claim A: "Murder is wrong because God says it is." 43.___
Claim B: "God says murder is wrong because it is."
The logical result of believing *either* Claim A or Claim B is that
I. some people think murder is right
II. it is difficult to know what God says
III. morality is logically independent of God's word
IV. atheists can be moral
V. religious people must use extreme self-discipline not to sin

The CORRECT answer is:
A. I *only*
B. I, V
C. II, IV
D. III, IV
E. II, IV, V

44. The philosopher(s) who claimed that religion was the opiate of the people is(are) 44.___
I. Aristotle
II. Plato
III. Marx
IV. Hegel
V. Kierkegaard

The CORRECT answer is:
A. I *only*
B. I, V
C. I, III
D. III *only*
E. V *only*

45. *Individualism* is a theory that is MOST incompatible with the philosophy of 45.___
I. Aristotle
II. John Stuart Mill
III. Jeremy Bentham
IV. Karl Marx
V. Nietzsche

The CORRECT answer is:
A. I *only*
B. I, II
C. II, III
D. I, IV
E. IV, V

46. *Organicism* is a theory of society that claims that 46.___
I. individuals make up the primary component of societies
II. societies are whole entities
III. the whole of society is greater than the sum of its parts
IV. humans are mutually independent of each other
V. humans are mutually dependent on each other

The CORRECT answer is:
A. I *only*
B. I, IV
C. II, III
D. II, III, IV
E. II, III, V

47. The philosopher(s) who believed that society should be classless is(are) 47.___
I. Aristotle
II. Plato
III. John Stuart Mill
IV. Karl Marx
V. Machiavelli

The CORRECT answer is:
A. I *only*
B. II *only*
C. I, III
D. II, IV
E. II, III, V

48. The philosopher(s) who believed that society should be ruled democratically is(are) 48.___
I. Aristotle
II. Plato
III. John Stuart Mill
IV. Hobbes
V. Machiavelli

The CORRECT answer is:
A. I *only*
B. I, II
C. II, IV
D. III *only*
E. III, IV, V

49. The philosopher(s) who believed that society ought to have an absolute monarch is(are) 49.___
I. Aristotle
II. Plato
III. Hobbes
IV. Machiavelli
V. Luther

The CORRECT answer is:
A. I *only*
B. I, II
C. II, IV
D. III, IV
E. III, IV, V

50. St. Thomas Aquinas' political philosophy emphasized 50.___
I. an egoist human nature
II. natural law
III. predestination
IV. free will
V. absolute rule

The CORRECT answer is:
A. I, V
B. II, III
C. II, IV
D. I, III
E. I, III, V

KEY (CORRECT ANSWERS)

1. E	11. C	21. D	31. C	41. C
2. C	12. D	22. D	32. E	42. C
3. D	13. D	23. E	33. D	43. D
4. C	14. D	24. A	34. B	44. D
5. D	15. B	25. D	35. E	45. D
6. C	16. C	26. D	36. D	46. E
7. B	17. E	27. C	37. C	47. D
8. C	18. C	28. D	38. D	48. D
9. D	19. C	29. D	39. E	49. E
10. C	20. D	30. D	40. D	50. C

TEST 2

DIRECTIONS: Each question or incomplete statement is followed by several suggested answers or completions. Select the one that BEST answers the question or completes the statement. *PRINT THE LETTER OF THE CORRECT ANSWER IN THE SPACE AT THE RIGHT.*

1. Machiavelli's political philosophy emphasized that 1.____
   I. the people know what is best for them
   II. rulers ought to appear just, but not really be just
   III. rulers who are just are the most successful
   IV. moral insight and sensitivity are more important than prudence
   V. ruling a society is based on strength and cunning

   The CORRECT answer is:
   A. I *only*
   B. I, IV
   C. II, III
   D. II, V
   E. I, II, IV

2. Terms associated with Machiavelli's political philosophy are 2.____
   I. morality
   II. glory
   III. self-interest
   IV. prudence
   V. power

   The CORRECT answer is:
   A. I, III
   B. II, V
   C. II, III, V
   D. III, V
   E. II, III, IV, V

3. Terms associated with Hobbes' political philosophy are 3.____
   I. morality
   II. glory
   III. self-interest
   IV. contract
   V. power

   The CORRECT answer is:
   A. I *only*
   B. I, IV
   C. II, III
   D. III, IV
   E. II, IV, V

4. *Theocracy* is a system of government in which _____ rule. 4.____
   I. aristocrats
   II. landowners
   III. rich people
   IV. the people
   V. religious leaders

   The CORRECT answer is:
   A. I *only*
   B. I, IV
   C. III, V
   D. II, III
   E. V *only*

5. Philosophers advocating a version of Social Contract Theory are 5.___
I. Machiavelli
II. Plato
III. Hobbes
IV. Locke
V. Mill

The CORRECT answer is:
A. I *only*
B. I, III
C. II, V
D. III, IV
E. II, IV, V

6. Philosophers who argued AGAINST a political philosophy based on contract theory are 6.___
I. Plato
II. Hobbes
III. Mill
IV. Rousseau
V. Marx

The CORRECT answer is:
A. I, III
B. II *only*
C. II, IV
D. I, III, IV
E. I, III, V

7. Philosophers who based their political philosophies upon a "state of nature" are 7.___
I. Plato
II. Locke
III. Hobbes
IV. Mill
V. Rousseau

The CORRECT answer is:
A. I, II
B. II, IV
C. II, III, V
D. I, III, IV
E. II, IV, V

8. Philosophers who postulate sociable and moral human beings in the state of nature are 8.___
I. Plato
II. Locke
III. Hobbes
IV. Mill
V. Rousseau

The CORRECT answer is:
A. I, II
B. II, V
C. I, III, IV
D. II, IV
E. II, IV, V

9. The political philosopher(s) credited with originating universal ethical egoism is(are) 9.___
I. Aristotle
II. Hobbes
III. Locke
IV. John Stuart Mill
V. Karl Marx

The CORRECT answer is:
A. I *only*
B. II *only*
C. III *only*
D. IV *only*
E. III, IV, V

10. The political philosopher(s) for whom the consent of the people is important is(are) 10.___

I. Plato
II. Hobbes
III. Mill
IV. Locke
V. Marx

The CORRECT answer is:
A. I *only*
B. I, V
C. II, III
D. IV, V
E. V *only*

11. Advocates of rule by aristocracy include 11.___

I. Burke and Paine
II. Aristotle and Burke
III. Plato and Luther
IV. Plato and Calvin
V. Burke and Calvin

The CORRECT answer is:
A. I *only*
B. II *only*
C. I, IV
D. II, III
E. II, V

12. The political philosopher(s) who believed that the sovereign provides SOLE support for social stability is(are) 12.___

I. Plato
II. Marx
III. John Stuart Mill
IV. Rousseau
V. Hobbes

The CORRECT answer is:
A. I *only*
B. I, III
C. IV *only*
D. II, III, IV
E. V *only*

13. The political philosopher(s) credited with instituting natural rights of life, liberty and the protection of property is(are) 13.___

I. Aquinas
II. Rousseau
III. Locke
IV. Hobbes
V. John Stuart Mill

The CORRECT answer is:
A. I *only*
B. I, II
C. III *only*
D. IV *only*
E. III, IV, V

14. The philosopher(s) who advocated disobedience to a sovereign who failed to protect the country from outside attack or to enforce contracts is(are) 14.___

I. Rousseau
II. Locke
III. Jeremy Bentham
IV. Hobbes
V. Karl Marx

The CORRECT answer is:
A. I, IV
B. IV *only*
C. III, IV
D. II, V
E. V *only*

15. The political philosopher(s) famous for basing proper government upon the "general will" is(are) 15.___
I. Rousseau
II. Locke
III. Hobbes
IV. Jeremy Bentham
V. Karl Marx

The CORRECT answer is:
A. I *only*
B. I, II
C. II, III
D. III
E. IV, V

16. The political philosopher(s) credited for originating liberalism is(are) 16.___
I. Plato
II. Marx
III. Bentham
IV. John Stuart Mill
V. Spinoza

The CORRECT answer is:
A. I *only*
B. I, V
C. II, III
D. III, IV
E. III, V

17. MAIN doctrines of Karl Marx's political philosophy include 17.___
I. individual sovereignty
II. general will
III. natural rights
IV. economic determinism
V. representative democracy

The CORRECT answer is:
A. I, V
B. II, III
C. IV *only*
D. III *only*
E. IV, V

18. The *primary* obstacle(s) to social justice, according to Marx, is(are) 18.___
I. wealth
II. private property
III. abundant labor
IV. scarce labor
V. poverty

The CORRECT answer is:
A. I *only*
B. II *only*
C. I, III
D. II, IV
E. V *only*

19. Karl Marx's political philosophy is BEST described as 19.___
I. dialectical idealism
II. material eclecticism
III. materialistic representation
IV. historical materialism
V. dialectical materialism

The CORRECT answer is:
A. I *only*
B. II, III
C. IV *only*
D. IV, V
E. II, IV

20. The historical precedent(s) for the American Bill of Rights was(were) 20.____
   I. Jeremy Bentham's hedonistic calculus
   II. John Stuart Mill's version of utilitarianism
   III. the 1689 English Bill of Rights
   IV. Rousseau's Social Contract Theory
   V. John Locke's natural rights theory

   The CORRECT answer is:
   A. I *only*
   B. II, IV
   C. III *only*
   D. IV, V
   E. V *only*

21. Thomas Jefferson borrowed *primarily* from the thoughts of the political philosopher 21.____
   A. Jeremy Bentham
   B. John Stuart Mill
   C. Jean-Jacques Rousseau
   D. John Locke
   E. Thomas Hobbes

22. The *primary* provision(s) of a Writ of Habeas Corpus is(are) 22.____
   I. a jury trial
   II. a trial by one's peers
   III. to protect citizens during civil war
   IV. to inform the accused of the charges against him
   V. to avoid retractive legislation

   The CORRECT answer is:
   A. I *only*
   B. I, II
   C. II, IV
   D. III, V
   E. IV *only*

23. The Bill of Rights in the United States is 23.____
   I. the first ten amendments to the Constitution
   II. the last ten amendments to the Constitution
   III. a direct result of Locke's political philosophy
   IV. derived from Rousseau's general will doctrine
   V. designed to protect citizens from the government

   The CORRECT answer is:
   A. I, III
   B. II, IV
   C. I, III, IV
   D. I, III, V
   E. II, III, IV

24. Of the several ideas developed by John Locke, our Declaration of Independence includes 24.____
   I. the right to bear arms
   II. equality of all human beings
   III. rights to property
   IV. the pursuit of happiness
   V. governmental obligation to protect citizens

   The CORRECT answer is:
   A. I, III
   B. II *only*
   C. II, III, IV
   D. I, IV, V
   E. II, III, IV, V

25. The philosopher(s) who developed a theory of history include(s) 25.___
I. Plato
II. Vico
III. Rousseau
IV. Marx
V. Hegel

The CORRECT answer is:
A. I *only*
B. I, II
C. II, III, V
D. II, IV, V
E. I, III, V

26. The philosopher(s) who believed that history is *primarily* the result of reason is(are) 26.___
I. Aristotle
II. Kant
III. Marx
IV. Hegel
V. Plato

The CORRECT answer is:
A. I, V
B. II, IV
C. III *only*
D. II, III
E. II, IV, V

27. The philosopher(s) whose theories are *dialectical* include 27.___
I. Plato
II. Aristotle
III. Hegel
IV. Marx
V. Locke

The CORRECT answer is:
A. I *only*
B. I, II
C. II, III, IV
D. I, III, IV
E. II, IV, V

28. The famous "Master-Slave Paradox" originated from the political philosophy of 28.___
I. Aristotle
II. Plato
III. Rousseau
IV. Marx
V. Hegel

The CORRECT answer is:
A. I, III
B. II, IV
C. III *only*
D. IV *only*
E. V *only*

29. The philosopher(s) whose theories are a direct reaction against Kant's is(are) 29.___
I. Hegel
II. Rousseau
III. Nietzsche
IV. Kierkegaard
V. Sartre

The CORRECT answer is:
A. I *only*
B. I, III
C. III, IV
D. III, V
E. III, IV, V

30. The philosopher(s) who considered aesthetics to triumph over morality was(were) 30.___
I. Plato
II. Aristotle
III. Kant
IV. Nietzsche
V. Mill

The CORRECT answer is:
A. I *only*
B. I, III
C. II, IV
D. IV *only*
E. IV, V

31. Friedrich Nietzsche authored 31.___
I. "The Fear and Trembling"
II. "The Genealogy of Morals"
III. "The Sickness Unto Death"
IV. "The Birth of Tragedy"
V. "The Critique of Pure Reason"

The CORRECT answer is:
A. I *only*
B. I, III
C. II, IV
D. II, V
E. V *only*

32. "Hedonism" differs from "Eudaemonism" in that the former means ____ while the latter means ____. 32.___
A. self-control; happiness
B. happiness; pleasure
C. pleasure; happiness
D. pleasure; asceticism
E. happiness; self-control

33. The philosophy of Francis Bacon 33.___
I. formed the substance of modern science
II. relied primarily upon common sense
III. emphasized critical observation
IV. rejected ultimate causes
V. followed closely upon Plato's

The CORRECT answer is:
A. I *only*
B. I, V
C. I, II, IV
D. I, III, V
E. I, III, IV

34. The philosopher(s) who largely contributed to the rise of modern science is(are) 34.___
I. Hume
II. Locke
III. Kant
IV. Rousseau
V. Bacon

The CORRECT answer is:
A. I *only*
B. I, III
C. I, II, IV
D. I, II, V
E. I, II, III

35. The philosophical school of *logical empiricism* includes the philosophers 35.___
I. James
II. Wittgenstein
III. Dewey
IV. Russell
V. Carnap

The CORRECT answer is:
A. I, III
B. II, IV
C. II, IV, V
D. III, IV
E. I, IV, V

36. A MAJOR division of philosophical schools in the modern era can be characterized as that between 36.___
I. reason and emotion
II. idealism and materialism
III. analytical and continental
IV. rational and existentialist
V. idealism and analytical

The CORRECT answer is:
A. I *only*
B. I, II
C. I, III
D. I, III, IV
E. I, III, V

37. The philosopher Lucretius wrote 37.___
I. "The Organon"
II. "On the Nature of Things"
III. "The Enneads"
IV. "The Meditations"
V. "The Nichomachean Ethics"

The CORRECT answer is:
A. I *only*
B. I, V
C. II *only*
D. III *only*
E. III, IV

38. Aristotle did NOT author 38.___
I. "The Organon"
II. "The Poetics"
III. "The Phaedo"
IV. "The Nichomachean Ethics"
V. "The Monologium"

The CORRECT answer is:
A. I, III
B. II, IV
C. III *only*
D. III, V
E. II, III, V

39. The author of the "Prolegomena to any Future Metaphysics" was 39.___
A. Spinoza
B. Kant
C. Descartes
D. Hegel
E. Rousseau

40. John Stuart Mill authored 40.___
I. "The Leviathan"
II. "Utilitarianism"
III. "An Introduction to the Principles of Morals and Legislation"
IV. "On Liberty"
V. "The Social Contract"

The CORRECT answer is:
A. I *only*
B. II *only*
C. I, III
D. II, IV
E. II, IV, V

41. The philosopher(s) who invented logical paradoxes is(are) 41.___
I. Zeno
II. Plato
III. Aristotle
IV. Buridan
V. Mill

The CORRECT answer is:
A. I *only*
B. II, III
C. I, IV
D. III, V
E. V *only*

42. Assumptions of modern game theory, used in moral and political philosophy, are 42.___
I. irrationality
II. rationality
III. choice
IV. maximization
V. non-cooperation

The CORRECT answer is:
A. I, III
B. II, IV
C. II, III, V
D. II, III, IV, V
E. All of the above

43. The existentialist philosopher(s) who was(were) also Marxist is(are) 43.___
I. Sartre
II. Kierkegaard
III. Nietzsche
IV. Camus
V. Marcel

The CORRECT answer is:
A. I *only*
B. II, III
C. III *only*
D. IV, V
E. V *only*

44. The existentialist philosopher(s) who was(were) NOT atheist is(are) 44.___
I. Sartre
II. Kierkegaard
III. Nietzsche
IV. Camus
V. Marcel

The CORRECT answer is:
A. I *only*
B. I, III
C. II *only*
D. II, IV
E. II, V

45. The BEST way to characterize individualist social theory, as opposed to collectivist theory, is that on the 45.___
I. former, individuals are mutually interdependent
II. former, individuals are not mutually interdependent
III. latter, the whole is equal to the sum of its parts
IV. latter, the whole is more than the sum of its parts
V. former, the whole is more than the sum of its parts

The CORRECT answer is:
A. I *only*
B. I, III
C. II, IV
D. I, III, V
E. II, III

46. *Collectivists* in political philosophy include 46.___
I. Plato
II. Mill
III. Hobbes
IV. Aristotle
V. Marx

The CORRECT answer is:
A. I *only*
B. V *only*
C. II, IV
D. I, IV
E. I, IV, V

47. The difference between *intrinsic* and *extrinsic* good is that 47.___
I. the former is not attainable
II. the former leads to nothing more
III. the latter leads to nothing more
IV. only one has been the object of philosophical scrutiny
V. one is a means and the other is an end

The CORRECT answer is:
A. I *only*
B. I, IV
C. II, V
D. III, IV
E. I, III, V

48. The philosopher(s) who believed that friends, money and social status are important include 48.___
I. Plato
II. Aristotle
III. Descartes
IV. John Stuart Mill
V. Marx

The CORRECT answer is:
A. I *only*
B. II *only*
C. I, III
D. II, V
E. II, IV, V

49. The term "teleological" comes from the Greek word "telos" meaning 49.___
I. love
II. purpose
III. goal
IV. logic
V. reason

The CORRECT answer is:
A. I *only*
B. I, II
C. II, III
D. I, IV, V
E. V *only*

50. According to Aristotle, a morally good person is one who 50.___
I. contemplates universal virtues
II. follows in Plato's path
III. leads an active and athletic life
IV. is healthy and good-looking
V. develops deliberate habits

The CORRECT answer is:
A. I
B. I, II
C. III, V
D. III, IV
E. III, IV, V

KEY (CORRECT ANSWERS)

1. D	11. E	21. D	31. C	41. C
2. E	12. E	22. E	32. D	42. E
3. D	13. C	23. D	33. E	43. A
4. E	14. B	24. E	34. D	44. E
5. D	15. A	25. D	35. C	45. C
6. E	16. D	26. B	36. D	46. E
7. C	17. C	27. D	37. C	47. C
8. B	18. B	28. E	38. D	48. E
9. B	19. D	29. E	39. B	49. C
10. D	20. C	30. D	40. D	50. E

EXAMINATION SECTION
TEST 1

DIRECTIONS: Each question or incomplete statement is followed by several suggested answers or completions. Select the one that BEST answers the question or completes the statement. *PRINT THE LETTER OF THE CORRECT ANSWER IN THE SPACE AT THE RIGHT.*

1. The term "philosophy" literally means 1.___
   A. abstract thinking
   B. love of wisdom
   C. speculative conjecture
   D. profound thought
   E. study of the human mind

2. Plato was the student of 2.___
   A. Aristotle
   B. Epictetus
   C. Homer
   D. Sophocles
   E. Socrates

3. Which of the following people did NOT discuss justice with Socrates in THE REPUBLIC? 3.___
   A. Glaucon
   B. Thrasymachus
   C. Epictetus
   D. Adeimantos
   E. Polemarchos

4. Epistemology is 4.___
   A. the Greek notion of excellence
   B. the study of ultimate reality
   C. the philosophy of language
   D. the theory of knowledge
   E. theology

5. In the CRITO, Plato reports that Socrates chose to stay in jail because 5.___
   A. Crito came too late
   B. Crito never spoke to Socrates
   C. Crito's argument was weak
   D. Plato did not pay the fine
   E. none of the above

6. Karl Marx, in his political theory, maintains that 6.___
   A. ideas are greatly dependent on economic conditions
   B. man is primarily a physical being
   C. belief in God is unfounded
   D. alienation is an effect of capitalism
   E. all of the above

7. Which philosopher maintained that we need to make a "leap of faith" to believe in God? 7.___
   A. Descartes
   B. St. Anselm
   C. Kierkegaard
   D. Kant
   E. Thomas Aquinas

8. The idea that God is the prime mover is associated with 8.___
   A. William James
   B. Karl Marx
   C. The Logical Positivists
   D. Aquinas
   E. two of the above

9. St. Anselm's argument for the existence of God contends that 9.___
   A. an examination of the universe confirms that God exists
   B. only priests can know for certain that God exists
   C. God's existence is only proved scientifically
   D. God's existence is entailed in the concept of "God"
   E. none of the above

10. Empiricism supports the view that 10.___
   A. there are innate ideas
   B. knowledge is unrelated to sense experience
   C. sense experience is the primary source of knowledge
   D. mysticism is the best way to know the world
   E. none of the above

11. The ontological argument maintains that the idea of a non-existent God is 11.___
   A. self-contradictory
   B. the belief of fools
   C. impious and heretical
   D. a consequence of not paying attention to the universe
   E. the work of the devil

12. The cosmological argument seeks to establish that 12.___
   A. the notion of God has little value
   B. God exists only in the mind of the believer
   C. the universe is the cause of God
   D. the universe and God are not related
   E. God is the cause of the universe

13. *A priori* and *a posteriori* are distinctions that refer to 13.___
   A. only ethical beliefs
   B. types of political theories
   C. the belief in God
   D. the philosophical contributions of Kant alone
   E. none of the above

14. Medieval philosophy is characterized by 14.___
   I. a great spirit of empirical research
   II. a rejection of empirical research
   III. a rejection of theology
   IV. a conviction that faith and reason are mutually exclusive
   V. theological questions

   The CORRECT answer is:
   A. I, III
   B. II, III
   C. II, IV, V
   D. III, IV, V
   E. all of the above

15. The Ontological Argument was advanced by 15.___
I. Schopenhauer
II. Nietzsche
III. St. Anselm of Canterbury
IV. Marcus Aurelius
V. St. Thomas Aquinas

The CORRECT answer is:
A. I *only*
B. II, III
C. II, IV
D. III, V
E. II, III, IV, V

16. "Act as if your action were to become a universal law of nature" is one of the possible expressions of 16.___
A. the categorical imperative
B. the hypothetical imperative
C. the principle of utility
D. an existentialist morality
E. none of the above

17. In general, teleological views hold that 17.___
A. acts are worthwhile only if done from a sense of duty
B. happiness is never pursued as a moral value
C. in ethics and science, matters can be understood in terms of their internal mechanisms
D. in ethics and science, matters can be understood in terms of their end result or purpose
E. none of the above

18. According to the concept of freedom found in Mill's ON LIBERTY, 18.___
A. the government should take an active role in redistributing income
B. the only grounds for interfering in the lives of adults is to prevent harm to others
C. men are equal only before the sight of God
D. because of historical changes, some men are justified in claiming greater liberty than others
E. all of the above

19. The categorical imperative states an ethical norm that 19.___
I. depends on certain sociological conditions
II. applies only when the person is interested in happiness
III. is derived from reason alone
IV. is universally valid
V. is only applicable in cases of emergency

The CORRECT answer is:
A. I, II
B. I, III
C. II, III, IV
D. III, IV
E. I, IV, V

20. John Locke believed that human beings were 20.___
A. born without rights
B. born with some rights which could be taken from them
C. born with inalienable rights
D. given their rights by society
E. able to earn their rights by obeying the laws of the state

21. A philosophical method called "methodological doubt" is associated with the philosophy of 21.___
    A. Plato
    B. Aristotle
    C. Descartes
    D. William James
    E. none of the above

22. Epistemology is a philosophical concern whose *main* interest is the clarification of problems related to 22.___
    A. moral values
    B. the nature and sources of knowledge
    C. the essence of the prime mover
    D. different logical questions
    E. a system of Eastern philosophy

23. Descartes reached the statement "I think, therefore I am" as a 23.___
    A. conclusion of his views of the solar system
    B. direct consequence of his own geometrical system
    C. conclusion of his methodological doubt
    D. direct consequence of his doubts about religion
    E. none of the above

24. Locke's concept of a "tabula rasa" (blank slate) states that the mind is 24.___
    A. born with innate ideas
    B. never able to know any truths
    C. able to understand the nature of God
    D. born with no innate ideas
    E. able to be reincarnated in animals

25. In the "Myth of the Cave," from Plato's REPUBLIC, the idea of the Good is symbolized by the 25.___
    A. cave itself
    B. fire in the cave
    C. sun
    D. shadows
    E. all of the above

26. In Plato's theory of value, 26.___
    A. values are dependent on personal preference
    B. man is the measure of all things
    C. it all depends upon where you were born
    D. there is a transcendent form of good
    E. sense experience is the key to the highest good

27. The Delphi Oracle was 27.___
    A. a philosophical system
    B. a contribution of the Milesian rationalists
    C. a figment of Plato's imagination
    D. a Greek religious institution
    E. one of Plato's dialogues

28. Greek Atomism was a system of cosmology developed by 28.___
    A. Aristotle
    B. Democritus
    C. Marcus Aurelius
    D. Thales
    E. none of the above

29. Cosmology is the study of the 29.___
   A. cosmos
   B. origin of the universe
   C. nature of the end of the world
   D. ultimate principles of reality
   E. all of the above

30. In "The Apology," Socrates *affirms* his 30.___
   A. support of atomism
   B. denial of life after death
   C. uncertainty about what death entails
   D. rejection of religion
   E. none of the above

31. *Platonism* is a philosophical style characterized by a(n) 31.___
   A. conviction that sense experience is primary
   B. denial of the possibility of knowledge
   C. contempt for mathematics
   D. affirmation of atomism
   E. rejection of sense experience

32. At his trial, Socrates was charged with 32.___
   A. a misuse of political power
   B. neglecting his military duties
   C. squandering public finances
   D. believing that water was the ultimate element
   E. corrupting the young

33. Pascal's Wager was a bet Pascal made 33.___
   A. with God
   B. with the devil
   C. concerning whether or not he should believe in God
   D. concerning whether or not he should live a virtuous life
   E. none of the above

34. The philosopher who wrote "Crito", "Phaedo", and "The Symposium" was 34.___
   A. Socrates
   B. Aeschylus
   C. Heraclitus
   D. Democritus
   E. none of the above

35. *Ethical absolutism* claims that 35.___
   I. when in Rome, do as the Romans do
   II. if something is wrong, it is always wrong
   III. if something is right, it is always right
   IV. right actions do not vary from culture to culture
   V. all of the above

   The CORRECT answer is:
   A. I *only*
   B. II *only*
   C. II, III
   D. II, III, IV
   E. V *only*

36. *Aesthetics* is that branch of philosophy that deals with 36.____
A. theological disputes
B. relativism in ethics
C. the nature of knowledge
D. the nature of reality
E. the nature of beauty

37. A person who claims to be an *agnostic* is one who doesn't know 37.____
A. what he/she is talking about
B. how to determine truth
C. whether or not God exists
D. whether or not anyone has knowledge
E. none of the above

38. An *atheist* is one who believes that 38.____
A. no one can know whether or not there is a God
B. there is a God
C. there is no God
D. it is irrelevant whether or not there is a God
E. philosophical knowledge is impossible

39. Algebra was invented by the philosopher 39.____
A. Maimonides
B. Aristotle
C. Plato
D. Descartes
E. Hobbes

40. *Altruism* is the view that states 40.____
A. no one does anything for another person, only for himself
B. everyone does things for others, occasionally
C. sometimes some people do things to benefit another
D. everyone acts only out of self-interest
E. helping others is bound to hurt yourself

41. A philosopher who argued persuasively AGAINST real altruism was 41.____
A. Plato
B. Aristotle
C. Kant
D. Bentham
E. Hobbes

42. An *analogy* is a 42.____
A. strange analysis revealing a hidden premise
B. simile
C. metaphor
D. comparison proving a point
E. none of the above

43. Analogies were a favorite technique of 43.____
A. Plato
B. Aristotle
C. Kant
D. Nietzsche
E. Moses Maimonides

44. An amoral person is one who 44.____
I. frequently commits wrong actions but doesn't know it
II. frequently commits right actions but doesn't know it
III. cannot tell the difference between right and wrong
IV. should be in prison
V. has failed to internalize moral values

The CORRECT answer is:
A. I, III
B. I, IV
C. II, III
D. III, IV
E. III, V

45. When a philosopher "analyzes" something, he 45.___
A. determines its validity
B. determines its soundness
C. takes it apart to examine its consequences
D. figures out what its consequences are
E. assembles various views to see what combinations work

46. When a philosopher performs philosophical "synthesis" on something, he 46.___
A. determines its validity
B. determines its soundness
C. takes it apart to examine its components
D. figures out what its consequences are
E. assembles various views to see what combinations work

47. Anarchists believe that 47.___
I. chaos is beautiful
II. governments are corrupt
III. the best government is one that governs least
IV. society can do without governments
V. humans are by nature socially cooperative

The CORRECT answer is:
A. I *only*
B. II
C. I, II, III
D. II, IV, V
E. III, IV, V

48. Anaximines thought the world consisted primarily of 48.___
A. water
B. fire
C. air
D. earth
E. gods

49. The Pre-Socratics were philosophers who 49.___
A. especially admired the Roman philosophers
B. especially despised the work of Aristotle
C. followed the paths laid out by Socrates
D. thought a lot but wrote little down
E. lived before Socrates

50. A skeptic is one who 50.___
A. believes nothing
B. believes many things but knows nothing
C. believes many things and knows that no one knows anything
D. believes everything
E. knows everything

KEY (CORRECT ANSWERS)

1. B	11. A	21. C	31. E	41. E
2. E	12. E	22. B	32. E	42. D
3. C	13. E	23. C	33. D	43. A
4. D	14. C	24. D	34. E	44. E
5. C	15. D	25. C	35. D	45. C
6. E	16. A	26. D	36. E	46. E
7. C	17. D	27. D	37. C	47. D
8. D	18. B	28. B	38. C	48. C
9. D	19. D	29. B	39. D	49. E
10. C	20. C	30. C	40. C	50. B

TEST 2

DIRECTIONS: Each question or incomplete statement is followed by several suggested answers or completions. Select the one that BEST answers the question or completes the statement. *PRINT THE LETTER OF THE CORRECT ANSWER IN THE SPACE AT THE RIGHT.*

1. A *cynic* is one who believes 1.___
   A. that everything in life works out for the best
   B. that everything in life works out for the worst
   C. that hope springs eternal
   D. in very little that cannot be scientifically proven
   E. that the natural lot of man is difficulty and unhappiness

2. *Anthropomorphism* is the view that 2.___
   I. there are many gods, not just one God
   II. no one really knows the nature of man
   III. all of nature behaves as humans do
   IV. humans lie at the center of the universe
   V. all of the above

   The CORRECT answer is:
   A. I *only*
   B. II *only*
   C. II, III
   D. II, IV
   E. III, IV

3. The FIRST sign of order in the universe, according to the ancients, was 3.___
   A. mathematics
   B. physics
   C. biology
   D. astronomy
   E. philosophy

4. St. Thomas became quite famous for revitalizing the views of 4.___
   A. Plato
   B. Socrates
   C. Epictetus
   D. Aristotle
   E. St. Anselm

5. *Thomistic* thinking emphasizes 5.___
   I. logic and order
   II. faith and emotion
   III. the natural purposes of things
   IV. religious piety
   V. none of the above

   The CORRECT answer is:
   A. I, III
   B. II, III
   C. III *only*
   D. III, IV
   E. V *only*

6. *Idealism* in philosophy is believing 6.___
I. that we live in the best of all possible worlds
II. that we live in the worst of all possible worlds
III. that truth can best be found in abstraction
IV. that truth can best be found in empirical sensation
V. in everything that can be argued for

The CORRECT answer is:
A. I *only*
B. III *only*
C. III, V
D. IV, V
E. V *only*

7. Statements that offer premises and a conclusion are called 7.___
A. conditionals
B. antecedents
C. consequents
D. arguments
E. modus ponens

8. *Asceticism* is the view that 8.___
I. the body is an evil, though a necessary one
II. people should have as many luxuries as possible
III. we eat to live
IV. we live to eat
V. excess is a flaw in human nature

The CORRECT answer is:
A. I, III, V
B. II, III
C. II, IV
D. III, V
E. IV, V

9. An *autonomous* person is one who 9.___
I. never listens to anyone else's opinions
II. always listens to people's opinions
III. remains independent in his/her actions
IV. decides for himself/herself
V. none of the above

The CORRECT answer is:
A. I *only*
B. II *only*
C. I, III
D. II, III
E. III, IV

10. *Axiology* is the study of 10.___
A. bodies
B. minds
C. values
D. moral principles
E. duties

11. Philosophers who were also scientists were 11.___
I. Aristotle
II. St. Thomas Aquinas
III. Sir Francis Bacon
IV. Socrates
V. Sir Isaac Newton

The CORRECT answer is:
A. I, II
B. I, III
C. II, IV
D. III, IV
E. III, V

12. The view that states that *only* sensory experience can confirm the existence of something is 12.___
I. idealism
II. materialism
III. empiricism
IV. verificationism
V. none of the above

The CORRECT answer is:
A. I, II
B. I, III, IV
C. II, III, IV
D. III, IV
E. V *only*

13. The philosophical and psychological school of *behaviorism* is represented by 13.___
I. Plato
II. Socrates
III. Skinner
IV. Freud
V. Watson

The CORRECT answer is:
A. II *only*
B. III *only*
C. III, IV
D. III, V
E. V *only*

14. The fundamental concept in the study of *ontology* is 14.___
A. truth
B. beauty
C. justice
D. being
E. nothingness

15. The philosopher who invented the "hedonistic calculus" was 15.___
A. John Stuart Mill
B. Plato
C. Epictetus
D. Jeremy Bentham
E. Thomas Hobbes

16. *Subjectivism* is the view that 16.___
I. external objects are dependent upon human consciousness
II. external objects exist independently of human consciousness
III. all knowledge is dependent upon human consciousness
IV. all knowledge is dependent upon empirical evidence
V. truth is invariable

The CORRECT answer is:
A. I, III
B. II, III
C. II, IV
D. II, IV, V
E. III, V

17. *Objectivism* is the view that 17.___
I. external objects are dependent upon human consciousness
II. external objects exist independently of human consciousness
III. all knowledge is dependent upon human consciousness
IV. all knowledge is dependent upon empirical evidence
V. truth is invariable

The CORRECT answer is:
A. I, III
B. II, III, V
C. II, IV, V
D. IV, V
E. V *only*

18. One of the MAJOR subjectivist philosophers was 18.____
A. Hume
B. Berkeley
C. Hobbes
D. Mill
E. Bentham

19. A term that can be applied to Aristotle, Democritus, Karl Marx, David Hume, and Hobbes is 19.____
A. idealist
B. materialist
C. empiricist
D. relativist
E. all of the above

20. The view that "the human mind is really only an illusion fostered by archaic thinking; in reality, we have only a brain" can be termed 20.____
I. empiricist
II. identity theory
III. materialist
IV. behaviorism
V. reductionism

The CORRECT answer is:
A. I, II, III
B. I, III, IV
C. III, IV
D. III, V
E. all of the above

21. The famous dictum, "Esse ist percipi" (if it is seen, it exists),is attributed to 21.____
A. Aristotle
B. Hume
C. Descartes
D. Berkeley
E. Locke

22. The view that humans have no "self," only a bundle of perceptions, is attributed to 22.____
A. Aristotle
B. Hobbes
C. Rousseau
D. Hume
E. none of the above

23. Under capitalism, 23.____
I. there is public ownership of the means of production
II. there is private ownership of the means of production
III. there is a gradual wearing away of the means of production
IV. profits fluctuate with market prices
V. profits do not fluctuate with market prices

The CORRECT answer is:
A. I, III
B. II, III
C. II, V
D. III, V
E. V *only*

24. The famous phrase "Cogito ergo sum" is attributed to 24.____
A. Plato
B. Berkeley
C. Moses Maimonides
D. Socrates
E. Descartes

25. The one thing Descartes said he knew *for certain* was that 25.___
 A. he doubted
 B. his consciousness existed
 C. only the mind could be known to exist
 D. to doubt is to think
 E. all of the above

26. *Dualism* is the view that 26.___
 A. all things come in twos
 B. dividing by two will always yield a mathematical result
 C. humans are made up of minds and bodies
 D. humans should emphasize their spiritual side, not their physical side
 E. all of the above

27. Dualism, historically speaking, can be attributed to the powerful thinking of 27.___
 A. Aristotle B. Plato
 C. Hume D. Kant
 E. Descartes

28. The term *catharsis* refers to the theory that 28.___
 A. humans become more aggressive watching hostilities on stage
 B. humans become less aggressive watching hostilities on stage
 C. humans remain constantly aggressive or not aggressive regardless of what they watch on the stage
 D. no one knows what can trigger aggressiveness
 E. all of the above

29. The theory of catharsis was first presented by artists in 29.___
 A. Greece B. Rome
 C. France D. Germany
 E. Britain

30. According to Karl Marx, history is the story of 30.___
 A. private accumulation of wealth
 B. greed
 C. warring tribes
 D. class struggle
 E. none of the above

31. A famous Roman philosopher and artist was 31.___
 A. Plato B. Cicero
 C. Marcus Aurelius D. Maimonides
 E. none of the above

32. Hellenic, as opposed to Dionysian art, emphasizes 32.___
 I. structure and order II. passion and emotion
 III. delicate shadings IV. clear lines
 V. none of the above

 The CORRECT answer is:
 A. I, IV B. II, III
 C. II, IV D. III, IV
 E. V *only*

33. *Cognition* refers to the process of 33.____
I. knowing
II. believing
III. thinking
IV. dreaming
V. wondering

The CORRECT answer is:
A. I *only*
B. I, II
C. I, III, V
D. II, IV, V
E. all of the above

34. Collectivist theories emphasize that 34.____
A. each person is out for himself
B. humans are much like other animals
C. humans are too greedy and unnecessarily cruel to be like other animals
D. humans are too cooperative and loving to be like other animals
E. the minimum component of society is the group

35. Conscience is heavily emphasized by 35.____
I. utilitarianism
II. deontology
III. existentialism
IV. theology
V. atheism

The CORRECT answer is:
A. II, III
B. II, IV
C. II, IV, V
D. III, IV
E. all of the above

36. A NECESSARY condition for *morality* is 36.____
A. legal systems
B. courts
C. police
D. choice
E. knowledge

37. Empirical knowledge can be challenged because we are sometimes 37.____
I. tricked by evil demons into believing what is not true
II. deceived into thinking we are awake when we are asleep
III. mistaken in what we think we see, as when we think the train tracks meet in the distance
IV. confused
V. all of the above

The CORRECT answer is:
A. I *only*
B. II *only*
C. II, III
D. III, IV
E. V *only*

38. Any philosophy, to be legitimate, should be 38.____
A. coherent
B. consistent
C. falsifiable
D. explainable
E. all of the above

39. *Contiguity* means that two things 39.___
A. are conspicuous
B. are caused one by the other
C. are caused by each other
D. resemble each other
E. are next to each other

40. The famous epistemological tract "Meditations" was written by 40.___
A. Anselm
B. Aquinas
C. Descartes
D. Maimonides
E. Aristotle

41. The well-known political philosophy in THE LEVIATHAN was authored by 41.___
A. Hume
B. Bentham
C. Mill
D. Hobbes
E. Adam Smith

42. The philosophers BEST known for studying and excelling in rhetoric are the 42.___
A. Greeks
B. Romans
C. Medievals
D. French
E. English

43. The correspondence theory of truth claims that statements 43.___
A. are true if they match up perfectly with reality
B. are true if they meet empirical standards of science
C. cannot be known to be since we don't know what's real
D. must be true independently of reality
E. none of the above

44. A *criterion* is a 44.___
A. method
B. system
C. style
D. standard
E. theory

45. The CRITIQUE OF PURE REASON was written by 45.___
A. Descartes
B. Plato
C. Aristotle
D. Plotinus
E. Kant

46. The philosopher and scientist famous for inventing evolutionary theory is 46.___
A. Aristotle
B. Francis Bacon
C. Charles Darwin
D. Isaac Newton
E. Alfred Einstein

47. *Deductive* arguments are arguments that 47.___
A. cannot be false
B. must be proven mathematically
C. do not rely upon empirical data
D. have conclusions that are certain
E. have conclusions that are only probable

48. *Inductive* arguments are arguments that 48.___
A. cannot be false
B. must be proven mathematically
C. do not rely upon empirical data
D. have conclusions that are certain
E. have conclusions that are probable

49. *Determinism* is the view that 49.___
I. physical existence admits of random events
II. every event has a cause
III. every effect has a cause
IV. we are determined by our environments
V. all of the above

The CORRECT answer is:
A. I, III, IV
B. II, III, IV
C. III, IV
D. IV *only*
E. V *only*

50. A *didactic* book is one that ____ the reader. 50.___
A. disillusions
B. teaches
C. disgusts
D. preaches at
E. entertains

KEY (CORRECT ANSWERS)

1. B	11. E	21. D	31. B	41. D
2. E	12. C	22. D	32. A	42. B
3. D	13. D	23. C	33. E	43. A
4. D	14. D	24. E	34. E	44. D
5. A	15. D	25. E	35. A	45. E
6. C	16. A	26. C	36. D	46. C
7. D	17. C	27. E	37. C	47. D
8. A	18. B	28. B	38. E	48. E
9. E	19. B	29. A	39. E	49. B
10. C	20. E	30. D	40. C	50. B

EXAMINATION SECTION

TEST 1

DIRECTIONS: Each question or incomplete statement is followed by several suggested answers or completions. Select the one that BEST answers the question or completes the statement. *PRINT THE LETTER OF THE CORRECT ANSWER IN THE SPACE AT THE RIGHT.*

1. According to Karl Marx, the MOST dangerous class of society is the 1.___
   A. bourgeoisie
   B. proletariat
   C. petit-bourgeoisie
   D. working class
   E. none of the above

2. The BEST solution to capitalism's problems, according to Marx, would be 2.___
   A. the abolition of private property
   B. a return of feudalism
   C. unions
   D. legal reform
   E. none of the above

3. According to Descartes, scientific knowledge is 3.___
   I. the best kind of knowledge we are able to get
   II. the worst kind of knowledge we are able to get
   III. apt to deceive us at best
   IV. imprecise
   V. possible to doubt

   The CORRECT answer is:
   A. I *only*
   B. II *only*
   C. II, III
   D. II, IV
   E. II, III, IV, V

4. Truths that Descartes thought were respectable were ones that 4.___
   I. were clear and distinct
   II. concerned people's existence
   III. concerned his own existence
   IV. concerned math and geometry
   V. all of the above

   The CORRECT answer is:
   A. I *only*
   B. I, II
   C. I, IV
   D. II, IV
   E. V *only*

5. David Hume proved that 5.___
   I. we have no good reason to believe that causation is a true principle of science
   II. experience is our only basis for knowledge
   III. ultimately there is no reason to assume the future will resemble the past

IV. we assume without evidence that like causes produce like effects
V. all of the above

The CORRECT answer is:
A. I, II
B. I, III, IV
C. II, III, IV
D. III, IV
E. V *only*

6. The thinkers who believed that humans by nature desire to know were 6.___
I. Plato
II. Aristotle
III. Skinner
IV. Freud
V. all of the above

The CORRECT answer is:
A. I, II, III
B. I, III, IV
C. II, IV
D. III, IV
E. V *only*

7. *Ethics* is the study of 7.___
A. principles of right conduct
B. standards of goodness
C. the ultimate goals of human society
D. human happiness
E. all of the above

8. The Greek word *polis* 8.___
I. means group
II. means city-state
III. means 'political'
IV. is the root of our word 'political'
V. all of the above

The CORRECT answer is:
A. I, IV
B. II, IV
C. III, IV
D. IV *only*
E. V *only*

9. The Greek word *psyche* 9.___
A. means soul
B. means mind
C. means wind
D. is the root of our word 'psychology'
E. all of the above

10. A knowledge of astronomy, studied by the earliest philosophers, gave early cultures 10.___
A. impressive educational systems
B. scholars who have lived down through history
C. naval power
D. indications of the scientific method
E. something to write books about

11. The point of the Socratic story of the "Ring of Gyges" is to show that 11.___
    A. everyone covets expensive royal jewelry
    B. anyone who has a valuable ring will use it for status
    C. anyone who could get away with it would exploit others
    D. no one with the Ring of Gyges would dare to be immoral
    E. all of the above

12. In an *aristocracy*, the rulers are 12.___
    A. those who want to be
    B. those with the most power
    C. those with the most wealth
    D. those with the best education
    E. all the people

13. In a *democracy*, the rulers are 13.___
    A. those who want to be
    B. those with the most power
    C. those with the most wealth
    D. those with the best education
    E. all the people

14. In an *oligarchy*, the rulers are 14.___
    A. those who want to be
    B. those with the most power
    C. those with the most wealth
    D. those with the best education
    E. all the people

15. In Plato's REPUBLIC, the rulers are 15.___
    A. men
    B. women
    C. men and women
    D. children
    E. philosophers

16. The 1600's were noteworthy for the 16.___
    I. rise in individualism
    II. rise of scientific investigation
    III. expansion of religious faith
    IV. origin of Protestantism
    V. all of the above

    The CORRECT answer is:
    A. I, II, IV
    B. II, III
    C. II, III, IV
    D. III *only*
    E. V *only*

17. An early forerunner of political philosophy was 17.___
    A. Hobbes
    B. Locke
    C. Plato
    D. Machiavelli
    E. Leibniz

18. Contract theorists include 18.___
    I. Marx
    II. Hobbes
    III. Locke
    IV. Rousseau
    V. John Stuart Mill

The CORRECT answer is:
A. I, II, III
B. II, III, IV
C. II, IV, V
D. III, IV, V
E. IV, V

19. The concept of free choice is *crucial* to the philosophical traditions of 19.____
I. Thomism
II. materialism
III. existentialism
IV. Kantianism
V. none of the above

The CORRECT answer is:
A. I, III, IV
B. I, IV
C. II, III, IV
D. III, IV
E. V *only*

20. The empiricist David Hume thought that ideas were 20.____
A. vague shadows of sensations
B. derived from empirical data
C. not as creative as they seem
D. caused by physical impressions
E. all of the above

21. Some concepts emphasized by Kant in his ethical theory were 21.____
I. consequences
II. motives
III. reason
IV. universal principles
V. all of the above

The CORRECT answer is:
A. I, II, III, IV
B. II, III, IV
C. III
D. IV *only*
E. V *only*

22. THE FEAR AND TREMBLING AND THE SICKNESS UNTO DEATH was written by 22.____
A. Sartre
B. Camus
C. Kierkegaard
D. Nietzsche
E. none of the above

23. Responsibility is a *fundamental* concept of 23.____
A. utilitarianism
B. Kantianism
C. existentialism
D. anarchism
E. none of the above

24. When Sartre said "Man is condemned to be free," he meant that 24.____
I. it is fun to make philosophical paradoxes
II. whether we like it or not, we must make choices
III. simply being born puts us in a situation of freedom
IV. even when we choose not to be free, we are free
V. to refuse to choose is to choose

The CORRECT answer is:
A. I, II
B. II *only*
C. II, III, V
D. II, IV, V
E. II, III, IV, V

25. The EXISTENTIALISTS believe that humans are by nature 25.___
A. greedy and selfish
B. loving and sociable
C. cooperative and friendly
D. free to make their own lives
E. none of the above

26. THE REBEL, THE MYTH OF SISYPHUS and THE STRANGER were written by 26.___
A. Sartre
B. Bergmann
C. Kierkegaard
D. Nietzsche
E. Camus

27. The philosophical school that believes life is irrational is 27.___
A. rationalism
B. idealism
C. materialism
D. existentialism
E. all of the above

28. Novelists who were early precursors of existentialism are 28.___
I. Tolstoy
II. Camus
III. Dostoevsky
IV. Balzac
V. Kafka

The CORRECT answer is:
A. I, III
B. II, III
C. II, III, V
D. III, IV
E. IV, V

29. *Quantitative* calculations are favored by the ethical school(s) of 29.___
I. hedonism
II. utilitarianism
III. egoism
IV. deontology
V. all of the above

The CORRECT answer is:
A. I, II, III
B. I, III, IV
C. II, III, IV
D. IV *only*
E. V *only*

30. In ethics, *cognitivism* means that 30.___
A. everyone can tell right from wrong
B. hardly anyone can tell right from wrong
C. knowing for certain what is right in each situation is mandatory to be a good person
D. moral judgments are true or false
E. all of the above

31. Happiness is the ultimate goal of 31.___
I. hedonism
II. utilitarianism
III. Aristotle's teleological ethics
IV. Plato's ethics
V. deontology

The CORRECT answer is:
A. I,II,III,IV
B. I,III,IV,V
C. II,III,IV
D. II,III,IV,V
E. III,V

32. The philosophers who divided up the soul into parts were 32.___
I. Plato
II. Aristotle
III. Hume
IV. Descartes
V. all of the above

The CORRECT answer is:
A. I, II
B. I, II, III
C. II, III
D. II, III, IV
E. V *only*

33. According to the ontological argument, God is 33.___
I. physically real
II. mentally real
III. perfect
IV. greater than which nothing can be conceived
V. loving

The CORRECT answer is:
A. I *only*
B. I, II, III
C. I, III, IV
D. III, IV
E. III, IV, V

34. According to Socrates, in the CRITO, the *ultimate* reason why Socrates is obligated to obey his death sentence is that he 34.___
A. loves Athens
B. is old anyway
C. was educated in public schools
D. never left his country
E. disliked bribery

35. INDUCTION is crucial to 35.___
A. theology
B. logic
C. science
D. computer science
E. all of the above

36. An ALLEGORY is a 36.___
A. parable
B. story that has at least two levels of interpretation
C. story with symbolic characters
D. narrative tale designed to prove a point
E. all of the above

37. Twentieth century philosophy is characterized by 37.___
I. logical analysis
II. sentential calculus
III. ordinary language analysis
IV. historical research
V. all of the above

The CORRECT answer is:
A. I, III
B. II, III
C. II, III, IV
D. III, IV
E. V *only*

38. The Delphi Oracle is really 38.___
A. the word of God
B. a fortune-telling device
C. the reason why Socrates got into trouble
D. a place one can still go see in Greece
E. all of the above

39. *Tacit consent* is the theory in political philosophy that states that 39.___
A. citizens sign contracts with governments
B. governments have the right to ask for loyalty oaths
C. citizens owe allegiance to the government ruling their chosen place of abode
D. citizens have no right to disobey a law they believe to be unjust
E. everyone has the right to plead the Fifth Amendment

40. One of the MOST socially revolutionary political philosophies was 40.___
A. deontology
B. intuitionism
C. utilitarianism
D. contract theory
E. Marxism

41. One of the MOST practical political philosophies for social reform was 41.___
A. deontology
B. intuitionism
C. utilitarianism
D. contract theory
E. Marxism

42. One reason that the subject of pain has fascinated philosophers for so long is that 42.___
I. everyone has experienced it
II. there is no empirical evidence that can be conclusively offered for it
III. it is intrinsically subjective
IV. it is intimately involved in epistemological questions
V. they are masochists

The CORRECT answer is:
A. I, III
B. II, III, IV
C. II, IV
D. III, IV
E. IV *only*

43. The *paradox of masochism* is that 43.___
I. only sadists would think about masochists
II. being mean to the masochist makes him happy
III. being nice to the masochist makes his suffer
IV. no one has a right to ask others to hurt them
V. all of the above

The CORRECT answer is:
A. I, II
B. I, II, III
C. II, III
D. II, III, IV
E. V *only*

44. The *invisible hand* theory was promoted by 44.___
A. Malthus
B. Jeremy Bentham
C. Karl Marx
D. Adam Smith
E. none of the above

45. The *first* philosopher to discuss DEMOGRAPHY was 45.___
A. Aristotle
B. Rousseau
C. Hobbes
D. Malthus
E. none of the above

46. THE ORIGIN OF THE FAMILY, PRIVATE PROPERTY AND THE STATE was written by 46.___
A. Rousseau
B. Marx
C. Locke
D. Engels
E. none of the above

47. According to Marxist theory, the family is 47.___
I. the only place where individuals can feel comfortable
II. the cellular form of class society
III. a form of natural oppression
IV. something to be abolished
V. all of the above

The CORRECT answer is:
A. I, II
B. II *only*
C. II, III
D. III, IV
E. V *only*

48. Jurisprudence is studied in 48.___
A. political philosophy
B. aesthetics
C. ethics
D. philosophy of law
E. none of the above

49. Rehabilitationism in theories of punishment is supported by 49.___
A. deontology
B. utilitarianism
C. existentialism
D. intuitionism
E. none of the above

50. One MUST be a moral agent to 50.___
A. act at all
B. be held responsible for actions
C. be legally culpable
D. do ethics
E. none of the above

KEY (CORRECT ANSWERS)

1. B	11. C	21. B	31. A	41. C
2. A	12. D	22. C	32. B	42. B
3. E	13. E	23. C	33. C	43. C
4. C	14. C	24. E	34. D	44. D
5. B	15. E	25. D	35. C	45. D
6. C	16. A	26. E	36. E	46. D
7. E	17. D	27. D	37. A	47. C
8. B	18. B	28. C	38. E	48. D
9. E	19. A	29. A	39. C	49. B
10. C	20. E	30. D	40. E	50. B

TEST 2

DIRECTIONS: Each question or incomplete statement is followed by several suggested answers or completions. Select the one that BEST answers the question or completes the statement. *PRINT THE LETTER OF THE CORRECT ANSWER IN THE SPACE AT THE RIGHT.*

1. People who are NOT moral agents include 1.___
   A. men
   B. women
   C. adults
   D. children
   E. foreigners

2. Retribution is a theory of punishment which maintains that 2.___
   A. those who are guilty deserve to be punished
   B. punishment serves as a good example to others
   C. the best deterrent is to punish wrongdoers
   D. morality is irrelevant to legal decision-making
   E. all of the above

3. When suicide was illegal in America, the rationale was that 3.___
   A. only stupid people would commit suicide
   B. only sick people would commit suicide
   C. no punishment could be harsh enough
   D. only God had the right to dispose of a life
   E. all of the above

4. The McNaghten Rule stipulated that 4.___
   I. certain people would be considered legally insane
   II. certain people would be allowed to sit on juries
   III. American law would differ drastically from English common law
   IV. being amoral is considered mitigatory
   V. none of the above

   The CORRECT answer is:
   A. I, II
   B. I, II, III
   C. I, IV
   D. II, III, IV
   E. V *only*

5. Philosophers who emphasized custom and habit are 5.___
   I. Plato
   II. Aristotle
   III. Hume
   IV. Rousseau
   V. Russell

   The CORRECT answer is:
   A. I, II, III
   B. I, II, IV
   C. II, III
   D. III *only*
   E. III, IV, V

6. A presupposition is 6.___
   A. necessary for an argument
   B. inevitable in ordinary discussion
   C. logically suspect
   D. an assumption
   E. none of the above

7. A statement can be DISPROVED by a counter-example if it is 7.___
   A. ridiculous
   B. unsupported
   C. a generalization
   D. a universalization
   E. none of the above

8. Historical materialism, the political philosophy, originated with 8.___
   A. Aristotle
   B. Plato
   C. Hobbes
   D. Marx
   E. none of the above

9. Epicurus, the hedonist, advocated 9.___
   I. pursuing knowledge
   II. collecting material possessions
   III. eating only bread and water
   IV. praying
   V. all of the above

   The CORRECT answer is:
   A. I, III
   B. I, III, IV
   C. II, III, IV
   D. III *only*
   E. V

10. The philosopher who complained of the "tyranny of the majority" was 10.___
   A. Hobbes
   B. Locke
   C. Mill
   D. Rousseau
   E. Marx

11. The MAIN concept in utilitarianism, that of *utility*, means 11.___
   I. usefulness
   II. futility
   III. prudence
   IV. happiness
   V. pleasure

   The CORRECT answer is:
   A. I, II, III
   B. I, IV
   C. II, IV
   D. III, IV, V
   E. IV, V

12. Philosophers who were *elitist* were 12.___
   I. Socrates
   II. Nietzsche
   III. John Stuart Mill
   IV. Kant
   V. Aristotle

   The CORRECT answer is:
   A. I *only*
   B. I, II
   C. I, II, III, IV
   D. I, II, V
   E. IV, V

13. In philosophy of law, *conventionalism* means 13.___
I. a person is immoral only if he/she breaks the law
II. a person does wrong only if there is a law against it
III. laws are based on nothing more than social convention
IV. laws are based on moral principles of society
V. none of the above

The CORRECT answer is:
A. I, III
B. II, III
C. II, IV
D. IV *only*
E. V *only*

14. EROS AND CIVILIZATION was written by 14.___
A. B. F. Skinner
B. Thoreau
C. Freud
D. Russell
E. none of the above

15. The theory of dualism generates the philosophical problems of 15.___
I. nominalism
II. epiphenomenalism
III. identity theory
IV. reductionism
V. none of the above

The CORRECT answer is:
A. I, II, III
B. II, III
C. II, III, IV
D. III, IV
E. V *only*

16. According to common sense realism, 16.___
I. people who doubt whether they exist are crazy
II. Cartesian doubt is absurd
III. objects are real whether we see them or not
IV. when we think we perceive something, and we have evidence, it is there
V. all of the above

The CORRECT answer is:
A. I, II, III, IV
B. II, III, IV
C. II, IV
D. III, IV
E. V *only*

17. *"The world exhibits order, therefore it was produced by an intelligent designer."* This argument is called the 17.___
A. ontological argument
B. cosmological argument
C. teleological argument
D. argument from faith
E. none of the above

18. The theological view that God inheres in all living creatures is 18.___
I. polytheism
II. monotheism
III. animism
IV. pantheism
V. all of the above

The CORRECT answer is:
A. I *only*
B. II *only*
C. II, III
D. III, IV
E. V *only*

19. Rule-utilitarianism claims that 19.___
   A. individual actions do not need to be morally evaluated
   B. the principle of utility is the proper criterion of right conduct
   C. act-utilitarianism is too time-consuming
   D. probabilities support moral rules
   E. all of the above

20. Act-utilitarianism claims that 20.___
   A. some actions are exceptions to moral rules
   B. the principle of utility is the proper criterion of right conduct
   C. rule-utilitarianism leads to bad consequences
   D. only acts which maximize future happiness ought to be done
   E. all of the above

21. THE PHILOSOPHY OF RIGHT was written by 21.___
   A. Marx
   B. Hegel
   C. Schiller
   D. Kant
   E. Pascal

22. Platonism holds that 22.___
   I. our senses give us reliable knowledge about the world
   II. there is a hidden, universal and unchanging world behind what we see
   III. all truth is universal
   IV. individuals are not significant overall
   V. all of the above

   The CORRECT answer is:
   A. I, II
   B. II, III
   C. II, III, IV
   D. II, IV
   E. V *only*

23. Philosophy of education concerns 23.___
   A. elucidation of proper didactics
   B. pedagogy
   C. propaganda
   D. political justification
   E. none of the above

24. Philosophy *primarily* deals with statements that are 24.___
   I. particulars
   II. generalizations
   III. universalizations
   IV. none of the above
   V. all of the above

   The CORRECT answer is:
   A. I, III
   B. II *only*
   C. II, III
   D. IV *only*
   E. V *only*

25. Some *fundamental* questions dealt with by MOST philosophers are 25.___

I. space
II. time
III. ideas
IV. death
V. all of the above

The CORRECT answer is:

A. I, II, III
B. I, III, IV
C. II, III
D. III, IV
E. V *only*

26. The proposition that the square of the hypotenuse equals the sum of the squares on the other two sides was discovered by 26.___

A. Descartes
B. Pythagoras
C. Leibniz
D. Aristotle
E. none of the above

27. Atomists believe that 27.___

A. the world is made up of combinations of the smallest particles
B. nuclear physics is the way of the future
C. the world is made up of water
D. earth, wind, fire and water make up reality
E. none of the above

28. Zeno's paradox tries to prove that 28.___

I. no one can get something from nothing
II. nothing can go through space because first it must go through half the distance, and half of that distance, etc.
III. space is infinitely divisible
IV. time is subjective
V. none of the above

The CORRECT answer is:

A. I, II
B. II *only*
C. II, III
D. III, IV
E. V *only*

29. A *deductive* argument which leads to a contradiction 29.___

I. shows that the premise is false
II. is the best kind of logic
III. characterizes the *reductio ad absurdum*
IV. is the standard type of syllogism
V. none of the above

The CORRECT answer is:

A. I, III
B. II, III
C. II, III, IV
D. IV *only*
E. V *only*

30. The psychologist who was NOT a physical determinist is 30.___
A. Skinner
B. Pavlov
C. Watson
D. Freud
E. none of the above

31. The philosopher who did NOT believe the soul is made up of distinct parts is 31.___
A. Plato
B. Socrates
C. Aristotle
D. Descartes
E. Hume

32. Platonism is characterized by 32.___
I. materialism
II. pantheism
III. empiricism
IV. idealism
V. universals

The CORRECT answer is:
A. I *only*
B. II *only*
C. II, III
D. III, IV, V
E. IV, V

33. Parmenides, with Socrates, argued the point that 33.___
A. life begins at death
B. existence is necessary
C. nothing cannot be something
D. everything is nothing
E. none of the above

34. In THE REPUBLIC, women are 34.___
I. treated as slaves
II. free to have children by any man
III. confined to childcare activities
IV. capable of being philosopher-kings
V. the equals of men

The CORRECT answer is:
A. I, II, IV
B. II, III, IV
C. II, IV
D. II, IV, V
E. IV, V

35. The ancient city-state whose politics inspired Plato's vision in THE REPUBLIC was 35.___
A. Athens
B. Selinus
C. Crete
D. Corinth
E. Sparta

36. An argument that can be repeated over and over, without reaching an end, is 36.___
A. a favorite of ancient philosophers
B. an infinite regress
C. a reductio ad absurdum
D. modus tollens
E. none of the above

37. The "Symposium," in which Socrates argues that true love requires wisdom in AT LEAST one of the lovers, is a(n) 37.___
A. Platonic dialogue
B. account of a class at the Academy
C. argument between Aristotle and Socrates
D. drinking party
E. all of the above

38. One of Socrates' famous claims was that 38.___
A. only my own consciousness can be proved
B. life begins at death
C. life is meaningless
D. wisdom is realizing one is ignorant
E. none of the above

39. A philosophical tract that is NOT written by Plato is 39.___
A. ORGANON
B. PHAEDO
C. SYMPOSIUM
D. MENO
E. REPUBLIC

40. One of the occupations NOT found in THE REPUBLIC is 40.___
A. guardian
B. worker
C. soldier
D. philosopher
E. musician

41. According to Socrates, the soul alternates between life in and out of a body. Therefore, 41.___
I. it is irrelevant what we learn in school
II. learning is remembering
III. death is no great tragedy
IV. man is immortal
V. none of the above

The CORRECT answer is:
A. I, III
B. II, III
C. II, III, IV
D. II, IV
E. V *only*

42. Given the argument offered by Socrates in the CRITO that each citizen is morally obligated to obey the law, Socrates would have to 42.___
I. agree with Martin Luther King that it is morally right to disobey unjust laws, like those prohibiting blacks to sit in the front of the bus
II. disagree with Martin Luther King, and say that King is morally obligated to sit in the back of the bus
III. disagree with King and suggest that if he does not like the laws in this country, he can leave
IV. agree with King that unjust laws have no right to be enforced
V. obey every law legitimately passed, even if it killed him

The CORRECT answer is:
A. I *only*
B. I, IV
C. II, III
D. II, III, V
E. III, V

43. The philosopher referred to as "The Stagyrite," and "The Philosopher" is 43.___
   A. Socrates B. Plato
   C. Aristotle D. Parmenides
   E. Pythagoras

44. Aristotle studied under 44.___
   A. Socrates B. Plato
   C. Parmenides D. Pythagoras
   E. Aeschylus

45. Aristotle's theories about ____ have been disproved. 45.___
   I. ethics II. politics
   III. biology IV. logic
   V. physics

   The CORRECT answer is:
   A. I, II B. I, III
   C. II, III D. III, IV, V
   E. III, V

46. A *primary* component of Aristotle's logic is 46.___
   A. matter B. form
   C. structure D. syllogism
   E. all of the above

47. *"All men are mortal. Socrates is a man. Therefore, Socrates is mortal."* 47.___
   This argument is
   I. invalid II. unsound
   III. syllogistic IV. valid
   V. sound

   The CORRECT answer is:
   A. I, II B. II *only*
   C. III, IV D. III, IV, V
   E. IV, V

48. Aristotle's logic is called 48.___
   I. propositional calculus II. Pythagorean mathematics
   III. theory of categories IV. theory of forms
   V. syllogistic

   The CORRECT answer is:
   A. I *only* B. I, III
   C. III, V D. IV, V
   E. V *only*

49. Philosophical schools advocating a withdrawal from public life are 49.___
   I. Cartesian rationalism II. Socratic idealism
   III. stoicism IV. cynicism
   V. utilitarianism

The CORRECT answer is:
A. I *only*
B. II *only*
C. II, III
D. III, IV
E. III, IV, V

50. Virtue, for Aristotle, was 50.___
I. the mean between the extremes
II. zeal over righteousness
III. impossible to attain
IV. incompatible with wealth and position
V. compatible with wealth and position

The CORRECT answer is:
A. I, V
B. II, III
C. II, IV
D. II, III, V
E. V *only*

KEY (CORRECT ANSWERS)

1. D	11. E	21. B	31. D	41. C
2. A	12. D	22. C	32. E	42. D
3. D	13. B	23. B	33. C	43. C
4. C	14. C	24. C	34. D	44. B
5. C	15. C	25. E	35. E	45. E
6. D	16. B	26. B	36. B	46. B
7. D	17. C	27. A	37. D	47. D
8. D	18. D	28. C	38. D	48. C
9. E	19. E	29. A	39. A	49. D
10. C	20. E	30. D	40. E	50. A

EXAMINATION SECTION

TEST 1

DIRECTIONS: Each question or incomplete statement is followed by several suggested answers or completions. Select the one that BEST answers the question or completes the statement. *PRINT THE LETTER OF THE CORRECT ANSWER IN THE SPACE AT THE RIGHT.*

1. Aristotle's theories are characterized *primarily* by their 1.____
   A. emphasis upon universal truths
   B. scientific accuracy
   C. emphasis upon teleology
   D. brevity
   E. all of the above

2. Euclid's geometry is characterized by 2.____
   I. inductive generalizations
   II. enumerative induction
   III. deduction
   IV. axiomatic precision
   V. all of the above

   The CORRECT answer is:
   A. I, II
   B. II, III
   C. III *only*
   D. III, IV
   E. V *only*

3. A Roman philosopher who taught Nero was 3.____
   A. Epictetus
   B. Chrysippus
   C. Zeno
   D. Marcus Aurelius
   E. Seneca

4. In his "Meditations," Marcus Aurelius advocates 4.____
   I. a withdrawal from life following stoicism
   II. an involvement in life as in Aristotle
   III. using power politics to achieve goals
   IV. using persuasion and argument to achieve goals
   V. separating private and public life

   The CORRECT answer is:
   A. I *only*
   B. II, IV
   C. III, IV
   D. III, IV, V
   E. IV, V

5. The philosopher who wrote the CONFESSIONS, in which he listed his many sins, and who later became a saint, is 5.____
   A. Descartes
   B. Aquinas
   C. Anselm
   D. Constantine
   E. Augustine

6. Medieval scholasticism is *closely* associated with the Greek philosopher 6.____
   A. Plato
   B. Aristotle
   C. Socrates
   D. Pythagoras
   E. Zeo

7. Thomism, in Catholic theology, originates with the philosopher 7.___
   A. Plato
   B. Anselm
   C. Aquinas
   D. Augustine
   E. none of the above

8. The *law of double effect* in Thomistic theology was designed to 8.___
   A. excuse an immoral action that is a necessary accompaniment of a good action
   B. show that only Catholics can be moral
   C. show the complexity of God's ways
   D. reveal the importance of consequences in moral thinking
   E. show that we can do two things at once

9. One of Aquinas' MAJOR philosophical accomplishments was 9.___
   A. showing the mystery of God
   B. resurrecting Platonic idealism in theology
   C. inventing mathematical systems
   D. proving that God created humans with free will
   E. proving that God has preordinated what humans will do

10. The philosopher who is still the *official* founder of Catholic theology is 10.___
   A. St. Augustine
   B. St. Aquinas
   C. St. Anselm
   D. Duns Scotus
   E. none of the above

11. Until St. Thomas Aquinas, Catholic theology disparaged 11.___
   A. faith
   B. revelation
   C. reason
   D. mystery
   E. women

12. Duns Scotus, disagreeing with Plato, said that 12.___
   A. will rules reason
   B. reason rules will
   C. logic rules thinking
   D. faith rules logic
   E. knowledge is based on universals

13. "Occam's Razor," proposed by William of Occam, is the view that 13.___
   A. whatever people believe is true
   B. truth is personal and not logical
   C. knowledge is based on experience
   D. simple explanations are always preferred
   E. complexity is the normal way of the world

14. Machiavelli wrote 14.___
   A. THE MEDITATIONS
   B. THE CONFESSIONS
   C. THE PRINCE
   D. THE REPUBLIC
   E. THE LEVIATHAN

15. One of the MOST important concepts in Machiavelli's political philosophy is 15.___
    A. prudence
    B. morality
    C. religion
    D. the papacy
    E. aristocracy

16. Two thinkers very much responsible for theological revolution in the period of the Enlightenment are ____ and ____. 16.___
    A. Descartes, Spinoza
    B. Leibniz, Descartes
    C. Luther, Calvin
    D. Calvin, Maimonides
    E. Erasmus, Maimonides

17. Copernicus was a philosopher who 17.___
    I. angered the Catholics and Lutherans both with his theory
    II. angered the Catholics but was praised by the Lutherans for his theory
    III. created a heliocentric hypothesis
    IV. challenged the omnipotence of God
    V. all of the above

    The CORRECT answer is:
    A. I *only*
    B. I, III
    C. II, III
    D. IV *only*
    E. V *only*

18. Philosophers who greatly hastened scientific knowledge are 18.___
    I. Newton
    II. Kepler
    III. Copernicus
    IV. Galileo
    V. all of the above

    The CORRECT answer is:
    A. I, II, III
    B. I, III, IV
    C. II, IV
    D. III, IV
    E. V *only*

19. The revolutionary idea advocated by Galileo, Copernicus, and Kepler was that the 19.___
    I. earth is not the center of the universe
    II. earth is the center of the universe
    III. planets orbit around the sun
    IV. sun is in orbit around the earth
    V. none of the above

    The CORRECT answer is:
    A. I *only*
    B. I, III
    C. II *only*
    D. II, IV
    E. V *only*

20. Galileo became famous for a scientific hypothesis 20.___
    I. concerning motions of bodies
    II. concerning static relationships of objects
    III. improving the theory of dynamics
    IV. proving that knowledge is impossible
    V. all of the above

The CORRECT answer is:
A. I, III
B. I, III, IV
C. II, III
D. II, III, IV
E. V *only*

21. A philosopher who profoundly influenced Thomas Hobbes' political philosophy is 21.___
A. Galileo
B. Aristotle
C. Plato
D. Augustine
E. Descartes

22. In 1640, Hobbes left England for France because of the publication of his book 22.___
A. THE CONFESSIONS
B. THE LEVIATHAN
C. THE PRINCE
D. THE REPUBLIC
E. THE MEDITATIONS

23. The *primary* focus of Hobbes' political theory was 23.___
A. sovereignty
B. peace
C. war
D. power
E. morality

24. For Hobbes, the WORST possible thing for a country is 24.___
A. civil society
B. the state of nature
C. corruption
D. civil war
E. foreign war

25. The view that the mental and the physical are separate is called 25.___
A. epiphenomenalism
B. empiricism
C. idealism
D. dualism
E. materialism

26. Spinoza wrote 26.___
I. THE ETHICS
II. THE LEVIATHAN
III. THE REPUBLIC
IV. THE TRACTATUS THEOLOGICO-POLITICUS
V. none of the above

The CORRECT answer is:
A. I *only*
B. I, IV
C. II, IV
D. III, IV
E. V *only*

27. Democracy was defended by 27.___
A. Plato
B. Aristotle
C. Machiavelli
D. Hobbes
E. Spinoza

28. THE MONADOLOGY was written by 28.___
A. Maimonides
B. Spinoza
C. Descartes
D. Leibniz
E. Hobbes

29. The country responsible for promoting empiricist philosophy the MOST was 29.___
   A. ancient Greece
   B. ancient Rome
   C. France
   D. Germany
   E. Britain

30. The *primary* British empiricists were 30.___
   A. Hobbes, Hume, Bacon
   B. Locke, Berkeley, Hume
   C. Locke, Leibniz, Spinoza
   D. Spinoza, Hobbes, Kant
   E. Bacon, Newton, Kepler

31. Empiricism holds that the mind is *originally* 31.___
   I. full of innate mechanisms
   II. comprised of innate ideas
   III. devoid of innate ideas
   IV. a blank sheet
   V. none of the above

   The CORRECT answer is:
   A. I, II
   B. I, III
   C. III, IV
   D. IV *only*
   E. V *only*

32. According to British empiricists, human thoughts are 32.___
   I. copied from sense experience
   II. vague shadows of physical objects
   III. clearer and more reliable than empirical data
   IV. dependent upon logical validity
   V. worthless

   The CORRECT answer is:
   A. I, II
   B. I, III
   C. II, III, IV
   D. III, IV, V
   E. V *only*

33. Doctrines of particular political importance gained from John Locke were 33.___
   I. private property
   II. divine right of kings
   III. natural rights
   IV. positive law
   V. none of the above

   The CORRECT answer is:
   A. I, II, III
   B. I, III
   C. III
   D. II, IV
   E. V *only*

34. The phrase *"Life, liberty and the pursuit of happiness"* 34.___
   I. was stolen from Hobbes' LEVIATHAN
   II. is derived from Locke's philosophy
   III. was changed from its original form, "Life, liberty, and the protection of property"
   IV. represents natural rights in liberal political theory
   V. all of the above

The CORRECT answer is:
A. I, II, III
B. I, III, IV
C. II, III, IV
D. II, IV
E. V *only*

35. The British empiricist philosophy LEAST compatible with common sense is that of 35.___
A. Locke
B. Hume
C. Berkeley
D. Hobbes
E. Bacon

36. The phrase that BEST describes Berkeley's epistemology is 36.___
A. British empiricism
B. logical positivism
C. existentialism
D. materialism
E. subjective idealism

37. Hume proved that we have NO evidence for believing in 37.___
I. cause and effect
II. human knowledge
III. love
IV. wisdom
V. all of the above

The CORRECT answer is:
A. I, II
B. I, II, III
C. II *only*
D. II, III, IV
E. V *only*

38. THE CONFESSIONS, THE SOCIAL CONTRACT and EMILE were *all* written by the philosopher 38.___
A. Descartes
B. Hobbes
C. Hume
D. Rousseau
E. Augustine

39. Kant's metaphysical system emphasized the distinctions between 39.___
I. noumena and phenomena
II. appearance and reality
III. love and hate
IV. reason and passion
V. things in themselves and things as they appear to us

The CORRECT answer is:
A. I, II, III
B. I, II, V
C. I, II, IV
D. II, V
E. IV, V

40. The MOST fundamental concept(s) in Kant's ethical system is(are) 40.___
A. benevolence
B. overall happiness
C. categorical imperative
D. consequences of our actions
E. all of the above

41. *"Don't vote for Mr. Smith in the upcoming election; he once failed an exam in high school, so he'd probably be unscrupulous."* 41.___
The fallacy in this argument is that of
A. provincialism
B. questionable cause
C. questionable classification
D. ad hominem
E. slippery slope

42. *"The universe, like a watch, must have a maker."* 42.___
The fallacy in this argument is that of
A. provincialism
B. questionable cause
C. questionable classification
D. questionable analogy
E. slippery slope

43. *"Her hair is so long and pretty; I'll bet she has thousands of dates."* 43.___
The fallacy in this comment is
A. provincialism
B. ad hominem
C. exaggeration
D. slippery slope
E. questionable analogy

44. *"Abortion is wrong because it is murder."* 44.___
The fallacy in this statement is
A. provincialism
B. questionable analogy
C. questionable classification
D. begging the question
E. slippery slope

45. *"Either those teachers care for the kids, and avoid a strike, or they strike and show they don't care for the kids."* 45.___
The fallacy in this argument is
A. provincialism
B. questionable analogy
C. false dilemma
D. straw man
E. inconsistency

46. *"Residents of Cleveland really don't want Mr. Smith for mayor; I asked 17 people in my neighborhood, and they think he shouldn't even run for office."* 46.___
The fallacy in this claim is
A. traditional wisdom
B. popularity
C. small sample
D. questionable classification
E. ambiguity

47. *"The General Surgeon has determined that breathing is dangerous to your health. This conclusion was drawn from a survey of 100 Canadian rats that have died within the past 5 years. All were habitual breathers."* 47.___
The fallacy in this argument is
A. traditional wisdom
B. popularity

C. questionable classification
D. questionable cause
E. questionable analogy

48. *"Of all the married women in the world, 73% wish they had married someone else."* 48.___
The fallacy in this claim is that of
A. suppressed evidence
B. faulty comparison
C. ambiguity
D. unknowable statistic
E. hasty conclusion

49. *"Freedom is wonderful, even if it makes some people feel imprisoned."* 49.___
The fallacy in this claim is that of
A. suppressed evidence
B. ambiguity
C. faulty comparison
D. unknowable statistics
E. hasty conclusion

50. *"Of course we ought to go there; Mr. Jones says so."* 50.___
The fallacy here is
A. suppressed evidence
B. fallacy of authority
C. unknowable statistics
D. hasty conclusion
E. traditional wisdom

KEY (CORRECT ANSWERS)

1. C	11. C	21. E	31. C	41. D
2. D	12. A	22. B	32. A	42. D
3. E	13. D	23. A	33. B	43. B
4. B	14. C	24. D	34. C	44. D
5. E	15. A	25. D	35. C	45. C
6. B	16. C	26. B	36. E	46. C
7. E	17. B	27. D	37. A	47. D
8. A	18. E	28. D	38. D	48. D
9. D	19. B	29. E	39. B	49. A
10. B	20. A	30. B	40. C	50. B

TEST 2

DIRECTIONS: Each question or incomplete statement is followed by several suggested answers or completions. Select the one that BEST answers the question or completes the statement. *PRINT THE LETTER OF THE CORRECT ANSWER IN THE SPACE AT THE RIGHT.*

1. One of G. E. Moore's MAIN claims as a common sense realist is that 1.___
   I. reality is actually much as it appears to us
   II. the fact that physics describes realitv differently from our ordinary way of doing so ,shows we are wrong
   III. theories that prove that what we see is not real are false
   IV. philosophical analysis can get carried away with itself
   V. all of the above

   The CORRECT answer is:
   A. I, II, III
   B. I, III, IV
   C. II, III
   D. II, III, IV
   E. V *only*

2. Ordinary language analysis is *fundamentally* a 2.___
   A. fallacious technique used by logicians
   B. method used by 17th century French thinkers
   C. 20th century philosophical method
   D. topic studied by linguists
   E. all of the above

3. *"Theories showing that I do not know I am holding this pencil in my hand right now MUST be false, BECAUSE they conclude this."* 3.___
   This argument, made by Moore, is
   I. a cheap argument to avoid the real issue
   II. question-begging
   III. a version of reductio ad absurdum
   IV. clever
   V. all of the above

   The CORRECT answer is:
   A. I *only*
   B. I, II
   C. II, III, IV
   D. III, IV
   E. V *only*

4. The theory of sense-data was a natural by-product of 4.___
   A. rationalism
   B. materialism
   C. empiricism
   D. nominalism
   E. realism

5. The chief problem with epistemological skepticism is that 5.____
   A. if it is true, no one knows anything
   B. if it is false, many philosophers have worked for nothing
   C. it is argued sloppily
   D. it is difficult to refute
   E. all of the above

6. One thing *challenged* by Humean skepticism is 6.____
   A. deriving a conclusion from enumerative induction
   B. our own existence
   C. God
   D. math
   E. none of the above

7. One *fundamental* concept in phenomenology is 7.____
   A. love
   B. death
   C. knowledge
   D. structuralism
   E. intentionality

8. Anarchism is a political philosophy espoused by ____ and ____. 8.____
   A. Karl Marx, Rousseau
   B. Thomas Hobbes, Locke
   C. Bakunin, Thoreau
   D. Stirner, Proudhon
   E. Emerson, Jefferson

9. Albert Camus was the author of 9.____
   I. BEING AND NOTHINGNESS
   II. THE MYTH OF SISYPHUS
   III. THE STRANGER
   IV. THE PLAGUE
   V. THE CRITIQUE OF DIALECTICAL REASON

   The CORRECT answer is:
   A. I, II
   B. II, III, IV
   C. III, IV
   D. III, V
   E. IV *only*

10. Jean-Paul Sartre was the author of 10.____
   I. BEING AND NOTHINGNESS
   II. THE MYTH OF SISYPHUS
   III. THE STRANGER
   IV. THE PLAGUE
   V. THE CRITIQUE OF DIALECTICAL REASON

   The CORRECT answer is:
   A. I *only*
   B. I, II
   C. I, IV, V
   D. I, V
   E. V *only*

11. Existentialism gets its name from its claims that 11.___
A. no one exists
B. solipsism is silly
C. emotions can prove things as well as reason
D. humans are helpless victims of human life
E. we do not choose to exist

12. Some typical claims made by existentialists are that 12.___
I. life is absurd
II. man's reason will save him from sin
III. God is a necessary component of human life
IV. truth is not universal but individual
V. all of the above

The CORRECT answer is:
A. I *only*
B. I, IV
C. II, IV
D. II, III, IV
E. V *only*

13. The theory existentialism challenged MOST was 13.___
A. nominalism
B. empiricism
C. rationalism
D. idealism
E. materialism

14.. Heidegger wrote the phenomenological and existential opus 14.___
A. BEING AND NOTHINGNESS
B. CRITIQUE OF DIALECTICAL REASON
C. CRITIQUE OF PURE REASON
D. BEING AND TIME
E. none of the above

15. Some MAJOR philosophers in the field of aesthetics are ____ and ____. 15.___
I. Croce, Coleridge
II. Dewey, Pater
III. Schiller, Tolstoy
IV. Hume, Kant
V. Descartes, Leibniz

The CORRECT answer is:
A. I, II, III
B. II, III, IV
C. III, IV
D. IV, V
E. V *only*

16. The existentialist philosopher famous for claiming that *"God is dead"* is 16.___
A. Sartre
B. Kierkegaard
C. Nietzsche
D. Heidegger
E. Camus

17. The existentialist philosopher used and distorted by Hitler was 17.___
A. Sartre
B. Kierkegaard
C. Nietzsche
D. Heidegger
E. Camus

18. The concept of the *ubermensch* was part of the existentialist philosophy of 18.___
A. Sartre
B. Kierkegaard
C. Nietzsche
D. Heidegger
E. Camus

19. Standard themes of existentialist writers are 19.___
I. love
II. death
III. sickness
IV. meaning
V. angst

The CORRECT answer is:
A. I, II
B. II, III
C. II, III, V
D. III, IV, V
E. all of the above

20. Correspondence theory states that 20.___
A. if you do not know how theories interrelate, you cannot achieve knowledge
B. statements that are logically valid correspond to rules of truth
C. statements that correspond to objective reality are true
D. reality corresponds with our minds
E. all of the above

21. The *primary* problem with correspondence theory is that 21.___
A. it assumes a metaphysics in order to judge the truth of statements
B. hardly any philosophers have argued for it
C. it is too simple and neat to be valid
D. no one really uses it
E. all of the above

22. Some of the MAJOR dilemmas posed by epistemology and metaphysics are that 22.___
I. both require at least a high school education to discuss
II. neither has achieved consensus in the philosophical community
III. we probably don't know anything *except* that we don't know anything
IV. each one presupposes the other
V. all of the above

The CORRECT answer is:
A. I, II
B. II, III
C. II, IV
D. III, IV
E. V *only*

23. Problems with the cosmological argument for the existence of God include that 23.___
I. it applies only to believers
II. it is patently false
III. there is so much evidence that it takes too much time to prove
IV. the evidence is not conclusive
V. there is evidence to the contrary

The CORRECT answer is:
A. I, II, III
B. II, III
C. II, IV
D. III, IV
E. IV, V

24. The interesting thing about first-person reports is that 24.___
A. they are not challenged by third-person reports
B. they claim epistemological priority
C. they cannot be proven false
D. to claim them is to verify them
E. all of the above

25. The *primary* method that is used to prove the existence of others is 25.___
A. empirical evidence
B. common knowledge
C. circumstantial evidence
D. argument by analogy
E. our intimacy with our mothers

26. Color is philosophically interesting because 26.___
A. it proves that some knowledge is gained empirically
B. we do not know if others see the same colors we do
C. objects do not have color
D. it is subjective
E. all of the above

27. Moore's common sense philosophy was really 27.___
I. a cop-out from the difficulties of metaphysics
II. a form of common sense realism
III. similar to some philosophers from Scotland
IV. compatible with the theories of Thomas Reid
V. an epistemological view

The CORRECT answer is:
A. I, II, III
B. I, II, III, IV
C. II, III, IV, V
D. III, IV, V
E. all of the above

28. The philosopher famous for having distinguished *knowledge by acquaintance* from *knowledge by description* is 28.___
A. Hume
B. Russell
C. Moore
D. Ayer
E. Hamlyn

29. Reductionism in 20th century philosophy is *usually* associated with 29.___
I. physicalism
II. determinism
III. idealism
IV. aesthetics
V. psychology

The CORRECT answer is:
A. I, II, V
B. II, III, V
C. II, IV
D. III, IV, V
E. all of the above

30. Logical atomism was a theory *primarily* promoted by 30.___
A. Moore
B. Russell
C. Ayer
D. Schlick
E. Hamlyn

31. The *fundamental* claim of LOGICAL ATOMISM is that 31.___
A. physics is mistaken
B. only quantum mechanics comes close to the truth
C. sense data are atomic particulars
D. the ultimate constituents of reality are atomic particulars
E. none of the above

32. Universities *usually* associated with ordinary language analysis are 32.___
I. Harvard
II. Princeton
III. Notre Dame
IV. Oxford
V. Cambridge

The CORRECT answer is:
A. I, II
B. II, III
C. III, IV, V
D. III, V
E. IV, V

33. An analytic truth 33.___
I. is tautological
II. cannot be false
III. is quite probably true
IV. is unknowable by humans
V. is true by definition

The CORRECT answer is:
A. I, II
B. I, II, III
C. II *only*
D. II, IV, V
E. II, V

34. An *a priori* truth 34.___
I. is tautological
II. cannot be false
III. is quite probably true
IV. is unknowable by humans
V. is true by definition

The CORRECT answer is:
A. I, II, IV, V
B. I, V
C. II, V
D. III, IV, V
E. IV, V

35. Some statements are 35.___
A. true analytically and a priori
B. true analytically but not a priori
C. false analytically
D. false a priori
E. all of the above

36. The difference between (a) "All effects are caused," and (b) "All events are caused" is that 36.___
A. (a) is true but (b) is false
B. (b) is false but (a) is true

C. (a) is analytic whereas (b) is a priori
D. (b) is analytic whereas (a) is a priori
E. (a) is analytic whereas (b) is synthetic and a priori

37. The 20th century philosopher who was NOT part of the ordinary language school of philosophy is 37.___
A. Ryle
B. Austin
C. Strawson
D. Dewey
E. Russell

38. The 20th century philosophical school of ordinary language analysis is *chiefly* characterized by its belief that 38.___
A. philosophers use too much jargon
B. the best language is the simplest
C. philosophical problems can be solved by analyzing language
D. the way we use words reveals philosophical assumptions
E. only English-speaking philosophers are legitimate

39. The skeptical claim, *"No one has any knowledge,"* is 39.___
A. false
B. true
C. paradoxical
D. fallaciously inferred
E. depressing

40. Some MAJOR problems of concern to 20th century Anglo-American philosophers are 40.___
A. love and death
B. being and nothingness
C. values
D. existence of other minds
E. none of the above

41. A synthetic statement 41.___
A. is also analytic
B. is also a priori
C. can be proved by definition
D. cannot be false
E. provides information

42. The philosophers who are transcendentalists in metaphysics are 42.___
I. Moore
II. Kant
III. Russell
IV. Hume
V. Locke

The CORRECT answer is:
A. I, II, III
B. II, III
C. II, III, IV
D. II, III, IV, V
E. III, IV, V

43. Philosophers who believe in transcendental metaphysics believe that ultimate reality 43.___
A. can be known *only* by God
B. is revealed through sensory experience
C. cannot be known without logic
D. is no different from ordinary perception
E. all of the above

44. One of the famous views propagated by Ludwig Wittgenstein is 44.___
A. common sense realism
B. that the best proof is to show necessary and sufficient conditions
C. that quantifying propositions increases philosophical knowledge by increasing precision
D. that it is impossible to define most words in our vocabularies
E. that there is a universal structure to language

45. Strawson, in discussing the existence of other persons, claims that 45.___
A. there is no evidence for the existence of others
B. there is only very shaky evidence for the existence of others
C. the concept of a person proves the existence of others
D. only persons worry about whether there are others
E. all of the above

46. The 20th century philosopher who emphasized Descartes' claim that *"I am not in my body like a pilot in a shop"* is 46.___
A. Russell
B. Strawson
C. Ryle
D. Moore
E. Wittgenstein

47. Problems with Cartesianism in epistemology are that 47.___
I. when we are unconscious we do not exist
II. it is at variance with our senses
III. it logically leads to solipsism
IV. it logically leads to skepticism
V. it confuses modern philosophers

The CORRECT answer is:
A. I, II, III, IV
B. I, III, IV
C. II, III, IV
D. III, IV, V
E. IV, V

48. Strawson claimed that 48.___
A. only love can break a heart
B. persons are self-ascribers of P-predicates
C. tautologies are not admissible as evidence of our existence
D. empiricism should be strictly applied in philosophical analysis
E. all of the above

49. Wittgenstein argued that private languages are 49.___
I. difficult to learn
II. impossible to imagine
III. conceptually confused
IV. out of fashion
V. useful for personal meditation

The CORRECT answer is:
A. I, II, III
B. II *only*
C. II, III
D. II, III, IV, V
E. III, V

50. The BEST evidence to use in discussing meaning, according to Wittgenstein, is 50.___
I. the dictionary
II. other people's opinions
III. how we behave when doing it
IV. how we refer to it in discussion
V. how we would teach it to others

The CORRECT answer is:
A. I *only*
B. II, III
C. II, IV, V
D. III, IV, V
E. IV, V

KEY (CORRECT ANSWERS)

1. B	11. E	21. A	31. D	41. E
2. C	12. B	22. C	32. E	42. B
3. D	13. C	23. E	33. A	43. C
4. C	14. D	24. E	34. B	44. D
5. D	15. A	25. D	35. E	45. C
6. A	16. C	26. E	36. E	46. B
7. E	17. C	27. C	37. D	47. A
8. C	18. C	28. B	38. C	48. B
9. B	19. E	29. A	39. C	49. C
10. D	20. C	30. B	40. D	50. D

EXAMINATION SECTION
TEST 1

DIRECTIONS: Each question or incomplete-statement is followed by several suggested answers or completions. Select the one that BEST answers the question or completes the statement. *PRINT THE LETTER OF THE CORRECT ANSWER IN THE SPACE AT THE RIGHT.*

1. The philosopher who claimed that *"I never can catch myself at any time without a perception"* is 1.___
   A. Hume
   B. Kant
   C. Russell
   D. Nietzsche
   E. Ryle

2. A *frequently* used concept in discussions of private languages is 2.___
   A. love
   B. fear
   C. worry
   D. doubt
   E. pain

3. One reason to DENY the existence of private languages is that 3.___
   A. doing so makes it easier to refute skepticism
   B. the existence of others is easier to prove without them
   C. it presupposes strict empiricism
   D. it creates more problems than it solves
   E. all of the above

4. Russell's opinion of ordinary language analysis was that 4.___
   I. it helped clear up common philosophical confusions
   II. it misleads philosophers into believing what they see
   III. metaphysical assumptions are built into language
   IV. only science can reveal reality, not language
   V. all of the above

   The CORRECT answer is:
   A. I, II, IV
   B. II, III, IV
   C. II, IV
   D. III, IV
   E. V *only*

5. The MOST notable advocate of logical atomism is 5.___
   A. Ayer
   B. Strawson
   C. Russell
   D. Moore
   E. Wittgenstein

6. Neutral monism claims that 6.___
   A. matter is neither good nor bad
   B. only one philosophical system can be true
   C. ethics is illegitimate
   D. there is only one form of matter, whether physical or mental
   E. none of the above

7. Logical positivism is a 20th century philosophy famous for 7.___
   A. challenging views on the basis of their falsity
   B. challenging views on the basis of their meaninglessness
   C. challenging views on the basis of their antiquity
   D. not challenging any views
   E. challenging views just to challenge

8. An everyday fact that proves transcendental metaphysics and argues *against* empiricism is that 8.___
   I. sticks in glasses of water are not really bent
   II. we frequently think we are awake when we are asleep
   III. only educated philosophers can do enough logic to prove metaphysics
   IV. ordinary people do not care about ultimate reality
   V. none of the above

   The CORRECT answer is:
   A. I, II
   B. I, II, III
   C. I, III, IV
   D. II, IV
   E. V *only*

9. A *central* concept in Strawson's metaphysics is 9.___
   A. neutral monism
   B. third person statements
   C. basic particulars
   D. logical fictions
   E. none of the above

10. Strawson and Russell DISAGREED, in their metaphysical theories, about whether or not 10.___
   I. reality was coherent
   II. reality was unified
   III. reality existed at all
   IV. we could ever know what is real
   V. philosophers earned a living

   The CORRECT answer is:
   A. I, II
   B. I, III
   C. II, IV
   D. II, IV, V
   E. IV, V

11. A *fundamental* criticism of utilitarian ethics is that it 11.___
   A. is inarticulately stated by its proponents
   B. is confusing to apply in practice
   C. requires moral agents to be unfair
   D. permits moral agents to be unfair
   E. is too narrow in its application

12. The four-step procedure recommended by utilitarian moral theory is: 12.___
   A. (1) plan your act; (2) become aware of your motives; (3) universalize; (4) act
   B. (1) read the Bible; (2) ask your parents; (3) flip a coin; (4) act
   C. (1) consider all alternatives; (2) calculate the consequences; (3) figure out utilities; (4) maximize happiness

D. (1) take a survey of utilities; (2) discount the utilities of anyone not present; (3) maximize the cardinal utilities involved; (4) act
E. (1) consider the moral rules in your theory; (2) decide whether the act involved is covered by a rule; (3) determine which rule; (4) act

13. Problems with the four-step procedure in the previous question are that it 13.___
I. is too time-consuming
II. is not useful for ignorant people
III. is too narrow in its application
IV. makes the same act right and wrong at different times
V. requires that one can read the future

The CORRECT answer is:
A. I, II, III, IV
B. I, II, IV, V
C. I, IV
D. I, IV, V
E. IV, V

14. Utilitarians defend their procedure by saying that 14.___
A. we use probabilities to predict the future
B. the best results can be obtained by using it
C. reason and calculation should be valued over common sense
D. what is right or wrong is not as obvious as it seems
E. all of the above

15. Deontology is *primarily* based upon the fact that 15.___
A. everyone already knows the difference between right and wrong
B. since no one knows what is right or wrong, it does not matter
C. some actions are wrong by their very nature
D. why people love calculations cannot be fathomed
E. prudence is irrelevant to ethics

16. Utilitarianism is a form of 16.___
I. metaphysics
II. consequentialism
III. teleological theory
IV. absolutism
V. nominalism

The CORRECT answer is:
A. I, II, III
B. II, III
C. II, III, IV
D. II, IV, V
E. III, IV, V

17. Deontology can be characterized as being 17.___
A. in opposition to consequentialism
B. in opposition to absolutism
C. relativistic
D. nominalistic
E. teleological

18. Things considered *important* by deontological ethics but *unimportant* by utilitarianism are ____ and ____. 18.____
    A. life, liberty
    B. love, death
    C. knowledge, education
    D. actions, consequences
    E. motives; intentions

19. The principle of Pareto Optimality in political theory claims that 19.____
    A. if a state is optimal for Pareto, it ought to be achieved
    B. optimality can be measured *only* by cardinal utilities
    C. optimality can be measured *only* by ordinary utilities
    D. optimality is achieved if no one can be made better off without making someone else worse off
    E. all of the above

20. One of the *fundamental* problems with Kant's version of deontology is that 20.____
    A. he stated it so obliquely and obtusely
    B. it rejects utilitarianism
    C. it is counter-intuitive
    D. it allows for no exceptions
    E. it gives no practical application

21. A version of deontology that is NOT absolutist is given by 21.____
    A. Ross
    B. Russell
    C. Moore
    D. Strawson
    E. none of the above

22. A *fundamental* concept in non-absolutist deontology is 22.____
    A. prima facie
    B. consequentialism
    C. practicality
    D. knowledge
    E. reason

23. Ethical systems that do NOT rely upon empirical calculation and investigation *usually* emphasize 23.____
    A. logic
    B. reason
    C. emotion
    D. intuition
    E. education

24. Decision-theoretic dilemmas assume that 24.____
    I. agents are altruistic
    II. agents are egoistic
    III. the choices are finite
    IV. agents are maximizers
    V. rationality is normal

    The CORRECT answer is:
    A. I, II, III, IV
    B. I, III, IV, V
    C. II, III, IV, V
    D. III, IV, V
    E. all of the above

25. John Rawls is famous for having written 25.____
    A. ANARCHY, STATE AND UTOPIA
    B. A THEORY OF JUSTICE
    C. PRINCIPIA MATHEMATICA
    D. THOUGHT
    E. REASON AND MORALITY

26. A *fundamental* concern in John Rawls' political philosophy is 26.___
   A. rights
   B. logic
   C. distribution of wealth
   D. the law
   E. none of the above

27. The MOST controversial principle introduced by Rawls is 27.___
   A. that equality of opportunity is primary
   B. the principle of universalization
   C. the principle of paternalism
   D. the difference principle
   E. none of the above

28. Rawls' recommendation for achieving justice DIFFERS from Pareto Optimality in that 28.___
   I. the former recommends that most people must be better off
   II. the latter recommends that most people must be better off
   III. only the former seeks to raise the worst off
   IV. only the latter seeks to lower the best off
   V. only the former seeks justice

   The CORRECT answer is:
   A. I, III
   B. II, III
   C. II, IV
   D. II, IV, V
   E. III, V

29. Robert Nozick's political philosophy claims to be 29.___
   A. valid and sound
   B. interesting
   C. inherited from Hume
   D. inherited from Locke
   E. inherited from Hobbes

30. The BEST way to characterize the logical positivist's approach to ethics is 30.___
   I. to avoid mentioning it
   II. as emotivism
   III. as non-cognitivist
   IV. scrupulous and conscientious
   V. all of the above

   The CORRECT answer is:
   A. I *only*
   B. II, III
   C. II, III, IV
   D. III, IV
   E. V *only*

31. G. E. Moore *chiefly* analyzed the concept of 31.___
   A. truth
   B. beauty
   C. justice
   D. wisdom
   E. goodness

32. Marx's theory of historical materialism is *frequently* said to be deterministic. This, however, CANNOT be, since 32.___
   A. Marx would have said so if he had meant it
   B. none of his statements lend themselves to that interpretation

C. he heavily emphasized the dialectical process
D. he never formally studied psychology
E. all of the above

33. One of the MOST important concepts in ancient Greek ethics is 33.___
A. noumena
B. Logos
C. anthropos
D. arete
E. akrasia

34. A popular theory, concerning the nature of justice, offered by Thrasymachus in the REPUBLIC, claims that 34.___
I. there is no such thing
II. justice is whatever the rulers say it is
III. might makes right
IV. everyone would be unfair if he thought he would be undetected
V. fair-minded people get exploited

The CORRECT answer is:
A. I, II, III
B. II, III, IV
C. II, IV, V
D. II, III, IV, V
E. III, IV, V

35. The beauty of the ontological argument is that, according to St. Anselm, NOT believing that God really exists 35.___
I. shows you haven't read the Bible
II. results in a contradiction
III. shows you don't even know what the word "God" means
IV. is complex and intricate
V. all of the above

The CORRECT answer is:
A. I, IV
B. II *only*
C. II, III
D. II, III, IV
E. V *only*

36. When confronted by a counter-example of altruism, the egoist denies the point that sometimes people act other than for self-interest by saying that 36.___
A. the agent is only trying to avoid feeling guilty
B. the person only wants to get to heaven
C. the feeling of satisfaction is what the individual wants
D. getting a good reputation is what is really desired
E. all of the above

37. A MAJOR problem with egoism is that it 37.___
A. is unfalsifiable
B. is devoid of information
C. distorts our ordinary way of speaking about altruism and self-interest
D. seems immoral
E. all of the above

38. Hobbes' *primary* opponent in the debate between altruism and egoism was 38.___
   A. Butler
   B. Hume
   C. Locke
   D. Bentham
   E. Mills

39. Techniques of logical analysis that are *frequently* used are 39.___
   I. algorithms
   II. predicate calculus
   III. Venn diagrams
   IV. truth tables
   V. calculators

   The CORRECT answer is:
   A. I, II, III
   B. II, III, IV
   C. III, IV
   D. III, V
   E. IV, V

40. Descartes, in his MEDITATIONS, proved that he knew for certain that 40.___
   I. his body existed
   II. his mind existed
   III. his body and his mind existed
   IV. he was a thinking thing
   V. all of the above

   The CORRECT answer is:
   A. I, II, IV
   B. II, IV
   C. III *only*
   D. III, IV
   E. V *only*

41. A MAJOR difference between Hobbes' State of Nature theory and Locke's is that 41.___
   I. in the former, there is no morality in the state of nature
   II. in the latter, there is morality in the state of nature
   III. Locke assumes altruism while Hobbes doesn't
   IV. benevolence is rejected in the latter
   V. all of the above

   The CORRECT answer is:
   A. I, II
   B. I, II, III
   C. I, III, IV
   D. II, IV
   E. V *only*

42. John Rawls' version of contract theory adds the feature of 42.___
   A. the difference principle
   B. altruistic agents
   C. the original position
   D. the hedonistic calculus
   E. none of the above

43. Hegel thought that the HIGHEST form of revealed Reason, in history, was 43.___
   A. God
   B. the state
   C. philosophy
   D. mathematics
   E. none of the above

44. The philosopher whose dialectic technique Marx "turned on its head" was 44.____
- A. Plato
- B. Aristotle
- C. Democritus
- D. Hegel
- E. Kant

45. *"If you drink a glass of wine tonight, tomorrow you'll want two glasses. The next day you'll probably want three, and before you know it, you'll be an alcoholic. Therefore, you shouldn't drink that glass of wine now."* 45.____
The fallacy in this argument is
- A. the naturalistic fallacy
- B. modus tollens
- C. non sequitur
- D. slippery slope
- E. none of the above

46. *"Since he punched me in the face for no reason, I have a right to punch him back."* 46.____
The fallacy here is
- A. the naturalistic fallacy
- B. modus tollens
- C. slippery slope
- D. non sequitur
- E. two wrongs don't make a right

47. *"I know I shouldn't have stolen the secrets, but everybody else does it."* 47.____
The fallacy in this argument is
- A. the naturalistic fallacy
- B. common practice
- C. non sequitur
- D. slippery slope
- E. modus tollens

48. *"This may not be the best thing to do, but it has always been done this way so far. Therefore, it ought to be done this way now."* 48.____
The fallacy in this argument is that of
- A. slippery slope
- B. non sequitur
- C. traditional wisdom
- D. ambiguity
- E. questionable analogy

49. *"This is a good thing to do because people in this area always do it this way."* 49.____
The fallacy here is that of
- A. ambiguity
- B. questionable cause
- C. tokenism
- D. provincialism
- E. guilt by association

50. *"Of course our company has equal opportunity; one of our top officers is Miss Jones, a black woman."* 50.____
The fallacy here is that of
- A. ambiguity
- B. questionable cause
- C. tokenism
- D. provincialism
- E. hasty conclusion

KEY (CORRECT ANSWERS)

1. A	11. D	21. A	31. E	41. A
2. E	12. C	22. A	32. C	42. C
3. E	13. D	23. D	33. B	43. B
4. B	14. E	24. C	34. D	44. D
5. C	15. C	25. B	35. C	45. D
6. D	16. B	26. C	36. E	46. E
7. B	17. A	27. D	37. E	47. B
8. A	18. E	28. E	38. A	48. C
9. C	19. D	29. D	39. C	49. D
10. A	20. D	30. B	40. B	50. C

TEST 2

DIRECTIONS: Each question or incomplete statement is followed by several suggested answers or completions. Select the one that BEST answers the question or completes the statement. *PRINT THE LETTER OF THE CORRECT ANSWER IN THE SPACE AT THE RIGHT.*

1. Social Darwinism claims that 1.___
   I. if you want something badly enough, you'll get it
   II. only the fit survive
   III. in human society, everyone is better off if the inferior ones are extinguished
   IV. everyone always acts in his own self-interest
   V. occasionally, altruism is possible

   The CORRECT answer is:
   A. I, II, III
   B. II, III
   C. II, III, IV
   D. II, III, V
   E. III, V

2. The flaws of Social Darwinism include 2.___
   I. a faulty analogy between animal and human society
   II. an historical precedence
   III. ambiguities in being "fit"
   IV. ambiguities in "extinguishing" unfit members of society
   V. there are no flaws

   The CORRECT answer is:
   A. I, III, IV
   B. I, IV
   C. II, III, IV
   D. II, IV
   E. V *only*

3. One of the social movements that follows as a logical consequence of Social Darwinism is 3.___
   A. genocide
   B. racism
   C. sexism
   D. eugenics
   E. utilitarianism

4. The model for Freud's conceptual scheme of the human mind originated with 4.___
   A. microbiology
   B. metaphysics
   C. hydraulics
   D. philosophy
   E. none of the above

5. Sartre's theory in BEING AND NOTHINGNESS can best be characterized as 5.___
   I. a variety of psychoanalysis
   II. obtuse and obscure
   III. dialectical social theory
   IV. serious, if misguided, speculation
   V. all of the above

The CORRECT answer is:
A. I, III
B. I, III, IV
C. II, III
D. III, IV
E. V *only*

6. Simone de Beauvoir, in her influential book THE SECOND SEX, draws upon the philosophical theory of 6.___
A. Kierkegaard
B. Camus
C. Nietzsche
D. Freud
E. Sartre

7. Nietzsche believed that 7.___
I. most people are followers, and only a few can lead
II. the rules of society apply only to followers, not to the strong
III. morality is an evil in itself
IV. only beauty is worth thinking about
V. all of the above

The CORRECT answer is:
A. I, II
B. I, II, III
C. II, III, IV
D. IV *only*
E. V *only*

8. Schopenhauer's philosophy has been considered a prime example of 8.___
A. optimism
B. pessimism
C. abstract idealism
D. empiricism
E. none of the above

9. Behavioral psychologists, as strict empiricists, are committed to the claim that 9.___
I. anything that cannot be empirically verified does not exist
II. only philosophical speculations can produce new knowledge
III. physical causes are the prime motivators of human behavior
IV. humans behave much as animals would in the same situation
V. whatever breathes can be motivated by fear and love

The CORRECT answer is:
A. I, II, III
B. I, III, IV
C. II, III, IV
D. III, IV, V
E. IV, V

10. Utilitarianism supports deterrence in punishment theory because the theory states that 10.___
A. guilty people should be forgiven out of charity
B. punishing people is one form of causing unhappiness
C. society does not benefit by retribution
D. guilt and innocence are too vague to be properly assessed
E. all of the above

11. The proviso in John Locke's theory that limits how much property a person can own claims that 11.___
I. any goods the individual cannot consume himself belong to others
II. property owners should not hoard goods until they spoil
III. a person can accumulate property, as long as there is as much left for others that is as good as what he has
IV. only if someone pays taxes to a government can he own property
V. all of the above

The CORRECT answer is:
A. I, II, III
B. I, III
C. II, III
D. II, III, IV
E. V *only*

12. The only two obligations the sovereign has, according to Hobbes, are to 12.___
A. supply prisons and schools
B. protect citizens from outside attack and provide educations
C. protect citizens from outside attack and collect taxes
D. protect citizens from outside attack and enforce contracts
E. enforce contracts and collect taxes

13. The *primary* justification for the government, according to Hobbes, is that 13.___
A. otherwise we would be miserable and unhappy
B. it provides social stability by making people keep their word
C. God has ordained that people be governed by wise rulers
D. otherwise thieves and murderers would go free
E. it provides a vindication agent

14. In CIVILIZATION AND ITS DISCONTENTS, Freud argues that 14.___
I. only through Eros does man get what he needs
II. love has many forms, including hatred
III. individuals must repress their desires for the sake of society
IV. society should sacrifice its interests so that individuals can fulfill themselves
V. there can be no harmony between humans individually and societies in general

The CORRECT answer is:
A. I, II
B. II, III
C. II, IV
D. III *only*
E. III, V

15. Marcuse, in EROS AND CIVILIZATION, disagrees with Freud, in CIVILIZATION AND ITS DISCONTENTS, by saying that 15.___
A. Freud's assumption of egoism and non-benevolence is false
B. Aristotle was correct in saying that man is a social animal

C. only through social activity can individuals fulfill themselves
D. capitalism requires egoistic psychology
E. all of the above

16. Civil disobedience requires acceptance of punishment because 16.___
I. otherwise the citizens disobeying the law are fomenting revolution
II. society wants to see rabblerousers put in jail
III. it is proof that those disobeying the law approve of the system in general
IV. Thoreau said so
V. all of the above

The CORRECT answer is:
A. I, III
B. II, III, IV
C. II, IV
D. III, IV
E. V *only*

17. G. E. Moore's "Naturalistic Fallacy" involved the claim that 17.___
A. anyone who adheres to nature's call is committing a fallacy
B. ethics has nothing to do with the animal kingdom
C. sociobiology is mistaken
D. what is good is not the same as what makes people happy
E. all of the above

18. The surplus labor theory of value claims that 18.___
I. capitalists like surpluses
II. labor is what makes the world go round
III. workers lose part of what they labor for
IV. people only have a right to the fruits of their own labor
V. all of the above

The CORRECT answer is:
A. I, II, III
B. II, III, IV
C. III, IV
D. IV *only*
E. V *only*

19. The difference between *normative* judgments and *descriptive* ones is that the 19.___
I. former are evaluative
II. latter are evaluative
III. former belong to science
IV. latter belong to science
V. distinction is important to ethics

The CORRECT answer is:
A. I, II, IV
B. I, IV, V
C. II, III, V
D. II, IV, V
E. IV, V

20. Of the many normative judgments one can make, such as ones including the term *good*, some are NOT moral judgments but rather _____ judgments. 20.___

I. aesthetic
II. practical
III. religious
IV. medical
V. functional

The CORRECT answer is:
A. I, II, III
B. I, III, V
C. II, IV, V
D. I, III, IV, V
E. all of the above

21. Methodological relativism claims that 21.___
A. though there may be right and wrong, we can never know what it is
B. there is no right and wrong
C. right and wrong depend on one's environment
D. moral judgments originate from customs
E. all of the above

22. Descriptive relativism claims that 22.___
I. customs vary, but this does not change what is right
II. customs vary from culture to culture
III. morality and custom have nothing in common
IV. mores vary but moral principles can still be true
V. none of the above

The CORRECT answer is:
A. I, III
B. II, IV
C. II, III, IV
D. IV *only*
E. V *only*

23. One crucial issue concerning the morality of abortion is 23.___
A. the separation of church and state
B. marriage laws
C. divorce statistics
D. the humanity of the fetus
E. government regulation

24. Natural rights are difficult to prove because 24.___
A. people disagree about which ones we have
B. separating what is natural from what is social is almost impossible
C. there is a history of Catholic theology behind natural rights
D. some rights said to be natural are culture-specific
E. all of the above

25. Positive law theories DISAGREE with natural law theories in that the 25.___
I. former consider official enactments as sufficient to establish law
II. former believe that law is value-neutral
III. latter believes that law is value-neutral
IV. latter believes that law is normative
V. former denies the obligatoriness of law

The CORRECT answer is:
A. I, II, IV
B. I, III
C. II, III
D. II, IV, V
E. IV, V

26. The view that laws are whatever the sovereign commands is 26.___
A. a natural law theory
B. a positive law theory
C. attributable to H. L. A. Hart
D. attributable to John Austin
E. none of the above

27. Rehabilitationists, in punishment theory, argue that 27.___
I. only deterrence is a valid consideration when contemplating punishing someone
II. only the guilty ought to be punished
III. whether someone deserves punishment or not is irrelevant
IV. punishment is a crime in itself
V. all of the above

The CORRECT answer is:
A. I, II, III, IV
B. I, IV
C. II, III
D. III, IV
E. V *only*

28. You are insane under the law if you 28.___
I. do not know the law
II. act very strangely in public
III. do not know what you are doing
IV. do not care what others think about you
V. do not know right from wrong

The CORRECT answer is:
A. I *only*
B. II, III
C. II, III, IV
D. III, IV
E. III, V

29. Lon Fuller, the philosopher of jurisprudence, DISAGREES with Hart in that 29.___
I. Fuller believes the law is normative while Hart does not
II. Hart believes the law is normative while Fuller does not
III. Fuller follows the natural law tradition
IV. Hart follows the natural law tradition
V. Fuller follows the positivist tradition

The CORRECT answer is:
A. I *only*
B. I, II
C. I, III
D. I, IV, V
E. III, V

30. THE CONCEPT OF LAW, the famous book of jurisprudential theory, was written by 30.___
A. Oliver Wendell Holmes
B. John Austin
C. H. L. A. Hart
D. Lon Fuller
E. none of the above

31. In philosophy of law, there is a dispute as to whether or not rules are 31.____

I. valid
II. sound
III. normative
IV. generalized
V. stipulative

The CORRECT answer is:

A. I, II, III, IV
B. II, IV, V
C. II, V
D. III, IV, V
E. IV, V

32. The *fundamental* reason for determining the legitimacy of law is that 32.____

A. it is coercive
B. otherwise we will not obey laws
C. judges require it
D. scholars love to do it
E. all of the above

33. John Stuart Mill's criterion for limiting the law's reach states that the law has no business with an individual's self-regarding actions. Problems with this include that 33.____

A. some individuals act destructively
B. it is difficult to act so as only to affect oneself
C. some individuals are self-destructive
D. police ignore Mill's criterion anyway
E. all of the above

34. One of the MOST important issues in philosophy of law is whether or not 34.____

A. we ought to have any laws at all
B. laws are morally binding
C. laws ought to legislate morality
D. people ought to be expected to know the law
E. we have enough laws

35. Problems with deterrence theory include 35.____

I. the fact that it does not work
II. the fact that it ignores what people deserve
III. its utilitarian foundation
IV. its humaneness
V. its lack of humaneness

The CORRECT answer is:

A. I, II, III, IV
B. I, II, V
C. I, III, V
D. II, III, V
E. II, IV

36. A MAJOR epistemological issue discussed in Plato's THEATETUS concerns 36.____

A. whether or not humans wish to have knowledge
B. whether or not knowledge is even possible
C. why it is difficult to discuss knowledge philosophically
D. the difference between knowledge and true opinion or belief
E. the nature of love

37. John Austin's "Descriptive Fallacy" states that 37.___
    A. we are mistaken if we think we can describe reality
    B. a person's claims to knowledge are always doubtful
    C. the claim "I know" does not describe anything
    D. the claim "I know" does describe something
    E. descriptions are always wrong

38. A MAJOR issue in theory of action is 38.___
    A. how many descriptions accurately apply to the "same" action
    B. whether or not we act at all
    C. whether or not intentions are necessary for action
    D. how we go from a thought to an action
    E. all of the above

39. Certain kinds of statements, according to John Austin, DO something, such as when "I promise" puts us under an obligation. These are 39.___
    A. exclamations
    B. interrogatives
    C. performatives
    D. indicatives
    E. subjunctives

40. In contemporary epistemology, an issue hotly debated is 40.___
    A. that concerning innate ideas
    B. whether or not God knows ahead of time what we will do
    C. whether or not we need to know metaphysics before epistemology
    D. whether knowledge is directly or indirectly given
    E. none of the above

41. The school of thought MOST OPPOSED to empiricism in epistemology is 41.___
    A. intuitionism
    B. physicalism
    C. determinism
    D. materialism
    E. none of the above

42. Universal ethical egoism can be stated as: 42.___
    A. Everyone ought to act in the self-interest of one agent
    B. I ought to act in my own self-interest
    C. Everyone ought to act in his own self-interest
    D. Everyone always acts in his own self-interest
    E. None of the above

43. An act that is *prima facie* wrong, such as killing, may be 43.___
    A. actually right
    B. still wrong
    C. unjustifiable
    D. justified
    E. none of the above

44. The principle in ethics that states that no one should make an exception of himself with respect to morality can be called a 44.___
    A. principle of fairness
    B. principle of goodness
    C. principle of universalization
    D. logical principle
    E. all of the above

45. *"A Nazi SS guard comes to the door of a German in 1939 and demands to know whether there are Jews hidden there. Mrs. X has hidden 3 Jews in her basement; is it right for her to lie?"* 45.___
The philosopher who believes Mrs. X should tell the truth is
A. Moore
B. Mill
C. Bentham
D. Kant
E. Sartre

46. The philosophical phenomenon of intentionality was originated by 48.___
A. Hegel
B. Marx
C. Sartre
D. Montague
E. Husserl

47. Epiphenomenalism is the theory in philosophy of mind that states that 47.___
A. the body and the mind both affect each other
B. bodies and minds never affect each other
C. we only really have a mind
D. we only really have a body
E. only the body affects the mind, not the reverse

48. One central issue in philosophy of mind is 48.___
A. analyzing love and hate
B. figuring out which psychological system is correct
C. determining what the "self" is
D. determining the nature of ultimate reality
E. analyzing epistemological criteria

49. "What is left over if I subtract the fact that my arm goes up from the fact that I raise my arm?" 49.___
This famous modern philosophical problem comes from
A. Russell
B. Strawson
C. Austin
D. Wittgenstein
E. Moore

50. Gilbert Ryle's "ghost in the machine" criticism challenged the view that 50.___
A. computers can think
B. volitions are necessary for actions
C. our thoughts are clear
D. most people think at all
E. all of the above

KEY (CORRECT ANSWERS)

1. B	11. A	21. A	31. D	41. A
2. A	12. D	22. B	32. A	42. C
3. D	13. B	23. D	33. B	43. D
4. C	14. E	24. E	34. C	44. C
5. A	15. E	25. A	35. B	45. D
6. E	16. A	26. D	36. D	46. E
7. A	17. D	27. B	37. C	47. E
8. B	18. C	28. E	38. E	48. C
9. B	19. B	29. C	39. C	49. D
10. B	20. E	30. C	40. D	50. B

EXAMINATION SECTION

TEST 1

DIRECTIONS: Each question or incomplete statement is followed by several suggested answers or completions. Select the one that BEST answers the question or completes the statement. *PRINT THE LETTER OF THE CORRECT ANSWER IN THE SPACE AT THE RIGHT.*

1. In Plato's APOLOGY, Socrates 1. ___
   I. apologizes to the people of Athens for his misbehavior
   II. defends himself against various charges
   III. participates in a typical Platonic dialogue
   IV. is found guilty
   V. All of the above

   The CORRECT answer is:
   A. I,II B. II,III C. III,IV D. II,IV E. V *only*

2. Socrates was not afraid of death because 2. ___
   I. he was satisfied with the worldly success he had already attained
   II. the soul is immortal
   III. he didn't know enough about it
   IV. he was sick of his poverty-stricken life
   V. None of the above

   The CORRECT answer is:
   A. I,II B. II,III C. III,IV D. IV *only* E. V *only*

3. In both the ancient and the modern sense of the word, "sophistry" means 3. ___
   A. manipulating arguments to prove a point
   B. accepting a fee for supplying arguments
   C. having knowledge or wisdom
   D. an occult science
   E. all of the above

4. The crimes Socrates was accused of included the following: 4. ___
   A. Being an atheist B. Creating his own gods
   C. Corrupting the youth
   D. Defeating good arguments with bad ones
   E. All of the above

5. The dialogues of Plato were 5. ___
   I. actually written by Socrates
   II. written by Plato but used all of Socrates' ideas
   III. dramatic inventions created by Plato alone
   IV. semi-historical accounts of conversations in Athens at the time
   V. unrelated to the political events of the time

   The CORRECT answer is:
   A. I,II,III B. III,V C. II,IV D. II,III,V E. I,IV

25. Logic is 25. ___
I. the study of a priori statements
II. something invented by Aristotle
III. concerned with concrete, empirical reality
IV. concerned with abstract relations
V. all of the above

The CORRECT answer is:
A. I,III B. I,II,IV C. I,II,III D. III,IV E. V *only*

26. A set of statements, one of which is a conclusion and the other(s) of which is a(are)premise(s), is a description of 26. ___
A. a syllogism B. an argument C. only valid arguments
D. only sound arguments E. all of the above

27. Epistemology concerns 27. ___
I. examining metaphysical claims to see whether they are true
II. the analysis of logical arguments
III. substantiating claims of knowledge
IV. physical facts
V. human beliefs

The CORRECT answer is:
A. I,III,V B. I,II,IV C. I,II,III D. III,V E. II,V

28. *"If p, then q; p, therefore q."* This is an example of 28. ___
I. Modus Ponens II. Modus Tollens
III. invalid reasoning IV. valid reasoning V. calculus

The CORRECT answer is:
A. I,II B. III,IV C. I,IV D. III,V E. I,V

29. The pantheist believes 29. ___
I. that there is only one God
II. that there are many gods
III. that both animate and inanimate nature are part of God's nature
IV. there is nowhere God cannot be found
V. all of the above

The CORRECT answer is:
A. I,III B. II,III C. III,IV D. I,II,III
E. V *only*

30. Some of the 10 terms of predication for Aristotle are: 30. ___
I. quantity and quality II. passivity and activity
III. place and position IV. time
V. none of the above

The CORRECT answer is:
A. I,III,V B. I,II,III,IV C. II,III,IV
D. I,IV E. V *only*

6. Crito tried to persuade Socrates to escape from prison because 6. ___
   A. people would think Plato abandoned his old friend Socrates
   B. there was plenty of money rrom foreigners to bribe the guards
   C. Socrates' verdict and sentence were unjust
   D. he had a moral obligation to his family
   E. All of the above

7. Socrates refused to escape from prison because 7. ___
   I. his act of disobedience would destroy the laws of the land
   II. he owed his life to the state
   III. he thought he would be caught
   IV. there was no other place to go
   V. by staying in Athens, he had tacitly agreed to obey the law

   The CORRECT answer is:
   A. I,II,V B. II,III,V C. I,III,IV D. III,IV,V
   E. I,II,III,V

8. Socrates was sentenced to die by means of 8. ___
   A. hanging B. being drawn and quartered C. drowning
   D. drinking hemlock E. being beheaded

9. Socrates' famous quotations include: 9. ___
   I. "Know thyself"
   II. "The unexamined life is not worth living"
   III. "I think, therefore I am"
   IV. "What is important is not to live, but to live well."
   V. "Everything that is true or real is eternal"

   The CORRECT answer is:
   A. I,II B. I,II,III C. I,II,III,IV D. I,II,IV,V
   E. All of the above

10. The Socratic method involves 10. ___
   A. challenging authorities B. playing devil's advocate
   C. asking questions D. defending unorthodox opinions
   E. close scrutiny of famous philosophical publications

11. In philosophy, an idealist is someone who 11. ___
   I. is optimistic about the future
   II. pursues abstract, not concrete, truth
   III. rejects the scientific method as unreliable
   IV. holds a metaphysical view opposed to materialism
   V. values ideas over physical objects

   The CORRECT answer is:
   A. I,II B. II,III C. I,II,III D. I,III,V E. II,III,IV,V

12. The jury at Socrates' trial was composed of 12. ___
I. twelve peers
II. over one hundred citizens of Athens
III. both men and women, young and old
IV. both friends and enemies
V. only propertied aristocrats

The CORRECT answer is:
A. II,III,IV B. I,III,V C. II,IV,V D. III,IV,V
E. I,III

13. Plato's famous THEORY OF FORMS concerned 13. ___
I. how citizens appeared to the rulers of the state
II. epistemological issues
III. universal truths
IV. Socrates' pursuit of the truth
V. all of the above

The CORRECT answer is:
A. I,II,III B. I,V C. I,IV,V D. II,III,V E. II,III,IV

14. Strictly speaking, the word "philosophy" means 14. ___
A. study of knowledge
B. the thought process
C. wise old men
D. love of wisdom
E. pondering profound problems

15. The Ontological Argument tries to prove 15. ___
I. that reality is not real
II. that ontology is the true root of metaphysics
III. that God exists
IV. that the meaning of terms is sufficient evidence for existence
V. none of the above

The CORRECT answer is:
A. I,II B. II,III C. III,IV D. IV,V E. I,V

16. The following statement, 16. ___

If there are more trees in the world than there are leaves on any one tree, and no trees have no leaves, then there are at least two trees with the same number of leaves,

is:
A. *False*, because it is highly improbable
B. *True*, because it is highly probable
C. *False*, because the last claim is unrelated to the first two claims
D. *True*, because 0=0
E. *True*, by definition

17. Empirical statements are 17. ___
I. a posteriori claims
II. provable through observable evidence
III. incapable of being 100% true
IV. never certain
V. essential to science

The CORRECT answer is:
A. I,II B. I,II,III C. I,IV,V D. I,II,III,IV
E. All of the above

18. A priori statements are 18. ___
    A. true (or false) by definition
    B. invariably 100% true or false
    C. provable through logical analysis
    D. reliant upon the meaning of terms
    E. all of the above

19. Descartes's famous claim, *"Cogito, ergo sum,"* means: 19. ___
    A. "Do unto others as you would have them do unto you"
    B. "I am, therefore I think."
    C. "I think, therefore I am."
    D. Thinking is the only true source of truth
    E. all of the above

20. Descrates's *"methodological doubt"* consisted of 20. ___
    A. wondering whether the chair he sat on was real
    B. wondering how we ever know whether we are dreaming
    C. casting off all claims that cannot be proved 100% true
    D. temporarily believing that life is an illusion
    E. all of the above

21. Some of the examples used by Descartes to justify his method include 21. ___
    A. hallucinations B. dreams C. balls of wax
    D. God may really be an Evil Demon E. all of the above

22. The Ontological Argument is 22. ___
    A. quite different from the Cosmological Argument
    B. originally the ingenious idea of a priest
    C. a good example of an a priori argument
    D. short, elegant, and abstract
    E. all of the above

23. Some of the flaws in the Ontological Argument are that 23. ___
    I. its crucial definition is ambiguous
    II. it assumes that existence is a predicate
    III. only religious people could possibly be persuaded by it
    IV. it requires a knowledge of mathematics to understand it
    V. a counter-argument refutes it easily

    The CORRECT answer is:
    A. I,III B. I,II,III C. II,III,IV D. III,IV
    E. I,II ,V

24. Metaphysics is 24. ___
    I. the study of ontology
    II. the basis of epistemology
    III. a field in philosophy that tries to answer the question, *"What is real?"*
    IV. an analysis of physics
    V. studied by all philosophers of any stature

    The CORRECT answer is:
    A. I,III B. I,IV C. II,IV D. IV,V E. III,IV

31. A deductive argument is one in which 31. ___
I. all the premises can be proved to be true
II. the conclusion is true if the premises are true
III. the proof lies in mathematical relations
IV. the conclusion it establishes is highly probable
V. the conclusion is either true or false

The CORRECT answer is:
A. I,III B. II,III C. I,II,III D. II,III,IV
E. II,III,V

32. Inductive arguments differ from deductive arguments in that 32. ___
I. the former establish a conclusion with certainty, not probability
II. the latter establish a conclusion with certainty, not probability
III. the former may have true premises and a false conclusion
IV. empirical evidence is relevant to the former and not to the latter
V. empirical evidence is relevant to the latter and not to the former

The CORRECT answer is:
A. II,III,IV B. I,II,III C. III,IV,V D. IV,V
E. II,III,V

33. The conclusion of an inductive argument 33. ___
A. may be established by several premises whose number is unlimited
B. is fortified by an increasing number of substantiating cases
C. is weakened by a lack of empirical evidence
D. is typically arrived at by enumerative induction
E. All of the above

34. A valid argument 34. ___
I. has true premises and a false conclusion
II. has true premises
III. has a conclusion that follows from the premises
IV. requires some experience with what it is about to prove it
V. is the same as a sound argument

The CORRECT answer is:
A. I,II B. III,IV C. III *only* D. III,V E. I,IV,V

35. A sound argument 35. ___
I. has true premises
II. has true premises and a false conclusion
III. is valid and has true premises
IV. has false premises and a true conclusion
V. always persuades an audience

The CORRECT answer is:
A. I,II B. II,IV C. III,V E. I,II,III D. I,III

36. *"There are no purple swans"* is an example of a(n) 36. ___
A. typical philosopher's puzzle
B. universalization
C. generalization
D. existential claim
E. negative existential universal

37. *"There are no purple swans"* is a 37. ___
A. statement that is obviously false
B. statement that one could prove, though with difficulty
C. statement that cannot be proved
D. good way to begin a conversation
E. statement that is obviously true

38. An analogy is a(n) 38. ___
A. literary metaphor
B. form of deductive argument
C. comparison used to prove a point
D. simile
E. All of the above

39. A non sequitur is a(n) 39. ___
I. statement that follows deductively from a premise
II. statement that does not follow deductively from a premise
III. statement that does not follow from the previous statement
IV. statement that contradicts itself
V. easy way to make people laugh

The CORRECT answer is:
A. I,II B. I,II,III C. III,IV D. III,V E. I,III,V

40. An *"equivocation"* is a fallacy in which 40. ___
I. the terms being used are vague
II. the terms being used are obscure and undefined
III. there is more than one meaning to one or more terms
IV. the central problem is invalidity
V. All of the above

The CORRECT answer is:
A. I *only* B. I,II,III C. IV,V D. III,V E. III,IV

41. If half the slots on a roulette wheel are red and the other half are black, and if 25 times in a row the ball falls in a red slot, *then* the chances 41. ___
I. are very high that the 26th ball will fall in a black slot
II. are better that the 26th ball will fall in a black slot
III. that the 26th ball will fall in a black slot are lower than they were
IV. that the 26th ball will fall in a black slot are 50-50
V. that the 26th ball will fall in a black slot are the same as for the previous 25 instances

The CORRECT answer is:
A. I,IV,V B. I,II,III C. IV,V D. III,IV E. I,V

42. Medieval philosophy was largely concerned with 42. ___
I. discussing the morality of torture devices
II. justifying the divine right of kings
III. theology
IV. studying Plato
V. translating Plato from ancient Greek into Latin

The CORRECT answer is:
A. I,II,III B. I,II,III,IV C. II,III,IV,V D. I,III,IV,V
E. III,IV,V

43. In all likelihood, ancient cultures *first* discovered order in the universe by 43. ___
I. developing complex mathematical systems
II. establishing stable political regimes that encouraged abstract research
III. observations made while hunting for food
IV. observations of the uniformity of the stars
V. researching religious documents

The CORRECT answer is:
A. I,II B. II,III C. III,IV D. IV,V E. II,IV

44. The following famous statement, *"All humans by nature desire to know,"* was *first* stated by 44. ___
A. Plato B. Saint Augustine C. Aristotle
D. Freud E. Democritus

45. A *reductio ad absurdum* argument is one that 45. ___
I. inductively proves what it sets out to prove
II. deductively proves what it sets out to prove
III. deductively proves the opposite of what it set out to prove
IV. succeeds only if it results in a condtradiction
V. is used frequently and very effectively

The CORRECT answer is:
A. I,II,III B. III,IV,V C. II,III,IV D. I,III,V
E. I,III

46. *"Social Darwinism"* is a theory about society developed by 46. ___
A. Freud B. Darwin C. John Stuart Mill
D. Herbert Spencer E. none of the above

47. *What* is TRUE of the following statement: *Human society benefits by the extinction of its unfit members?"* It 47. ___
I. *relies upon* an analogy between biological survival and social relations
II. *accurately describes* current social systems
III. *presupposes* that what is biologically good is morally good
IV. *confuses* natural laws with moral laws
V. *justifies* continued exploitation of the sick and the poor

The CORRECT answer is:
A. I,III,V B. I,II,III,IV C. II,IV,V
D. I,III,IV,V E. II,III,IV,V

48. The *fundamental* issue in political philosophy is 48. ___
A. justice
B. governmental regulations
C. the relationship of the church and the state
D. courts and the law
E. democratic theories of representation

49. *First and foremost,* social justice concerns 49. ___
A. equal opportunity B. reasonable legislation
C. a good court system
D. the distribution of economic wealth
E. formal rules of election

50. A tautology is a(n) 50. ___
I. contradictory statement
II. invalid argument
III. true statement
IV. a priori statement
V. statement that is true by definition

The CORRECT answer is:
A. III,IV,V B. I,II,III C. I,III,V D. II,IV,V
E. None of the above

KEY (CORRECT ANSWERS)

1.	D	11.	E	21.	E	31.	E	41.	C
2.	B	12.	C	22.	E	32.	A	42.	C
3.	A	13.	E	23.	E	33.	E	43.	E
4.	E	14.	D	24.	A	34.	B	44.	C
5.	C	15.	C	25.	B	35.	D	45.	B
6.	E	16.	E	26.	B	36.	E	46.	D
7.	A	17.	E	27.	D	37.	C	47.	D
8.	D	18.	E	28.	C	38.	C	48.	A
9.	D	19.	E	29.	C	39.	D	49.	D
10.	C	20.	C	30.	B	40.	E	50.	A

TEST 2

DIRECTIONS: Each question or incomplete statement is followed by several suggested answers or completions. Select the one that BEST answers the question or completes the statement. *PRINT THE LETTER OF THE CORRECT ANSWER IN THE SPACE AT THE RIGHT.*

1. Machiavelli is credited with 1. ___
   A. developing the first theory of political science
   B. establishing the ways and means of obtaining political power
   C. using his theoretical knowledge to gain a position of political power
   D. challenging religious values
   E. All of the above

2. In THE PRINCE, Machiavelli *basically* 2. ___
   I. formulates abstract principles of political philosophy
   II. uses anecdotes from Italian history to prove his point
   III. tries to show how valuable political experience is to members of the nobility
   IV. draws upon the accepted theories of his day to draw attention to himself
   V. offers advice to would-be rulers

   The CORRECT combination is:
   A. I,II,V B. I,IV,V C. I,II,III D. II,III
   E. II,III,V

3. Some of the *main* points Machiavelli makes are: 3. ___
   I. Members of society cannot tolerate freedom
   II. Rulers do not need to appear honest or generous; all they need is power
   III. Political power depends primarily upon talent and ability in a ruler
   IV. Brutality and ruthlessness are sometimes justified if done for the benefit of society
   V. People seek to maximize their own self-interest

   The CORRECT combination is:
   A. I,II,III B. III,IV, V C. II,III,V D. I,III,IV
   E. II,III,IV

4. If Machiavelli's advice could be captured in one word, it would be 4. ___
   A. power B. control C. fortune D. ability
   E. prudence

5. Some scholars have challenged Machiavelli's status as the first political scientist on the grounds that: 5. ___
   I. His morality was simply too repulsive to gain academic respectability
   II. No scholarly work done before 1920 can possibly be legitimate
   III. Historical anecdotes are no substitute for statistical research
   IV. He changes his principles frequently, depending upon the situation he is discussing
   V. His primary purpose in writing THE PRINCE was to get a job

   The CORRECT combination is:
   A. I,II B. I,II,III C. I,II,III,IV D. I,III,V
   E. All of the above

6. Thomas Hobbes' *primary* purpose in his book THE LEVIATHAN (1651) was 6. ___
   A. to show how human nature and society fulfilled God's wishes
   B. to prove that human nature is basically egoistic
   C. to illustrate how mechanistic psychology can help substantiate political theory
   D. to justify the citizen's obligation to the state
   E. All of the above

7. In Hobbes' *"state of nature"* theory, humans are, without laws and government, 7. ___
   I. energetic and cooperative
   II. envious and acquisitive
   III. selfish and greedy
   IV. lonely and insecure
   V. equal in their ability to kill one another

   The CORRECT combination is:
   A. II,III,IV,V B. I,II,III,IV C. I,III,IV,V
   D. I,II,III,V E. All of the above

8. Egoism is a theory that states 8. ___
   I. everyone *always* seeks to maximize his/her self-interest
   II. everyone *usually* seeks to maximize his/her self-interest
   III. everyone seeks his/her own interest, whether or not it is at another's expense
   IV. everyone seeks to increase his own interest primarily to thwart others
   V. any apparently altruistic act is really self-interested

   The CORRECT combination is:
   A. I,II B. I,V C. I,III,V D. I,II,III E. II,III,IV

9. Altruism is the theory that states that 9. ___
I. everyone always seeks to maximize others' interest
II. everyone seeks to act in his/her own interest as well as in others'
III. not all actions are done in order to maximize the self-interest of the agent
IV. egoism and altruism are not mutually exclusive
V. egoism is false

The CORRECT combination is:
A. I,II,III,IV B. II,III,IV,V C. I,III,V D. II,III,IV
E. All of the above

10. Since life in the *"state of nature,"* according to Hobbes, is *"nasty, brutish, and short,"* a government is necessary in order to 10. ___
A. monopolize coercive power into one agency
B. enforce agreements and contracts
C. defend the country against outside attack
D. give people a reason not to maximize self-interest
E. All of the above

11. Hobbes is a nominalist because 11. ___
I. he claims that people are egoistic
II. he uses scientific reasoning to establish his conclusions
III. he believes that there is no justice or goodness in nature
IV. only the state can create rights
V. past experience is not a guide for judging actions

The CORRECT combination is:
A. I,III,IV B. I,II,III C. II,IV D. III,IV
E. IV,V

12. A hedonist is one who believes that 12. ___
A. love makes the world go round
B. pleasure is the highest good
C. only power brings success
D. truth is always more persuasive than falsity
E. None of the above

13. Physicalism is a 13. ___
I. variety of materialism
II. variety of idealism
III. theory that identifies causes and physical objects
IV. view that reduces psychology to physiology
V. theory that claims that every effect has a cause

The CORRECT combination is:
A. I,II B. I,II,III C. I,III,IV,V D. II,III,IV,V
E. All of the above

14. Empiricism is a theory that states that 14. ___
I. anything that cannot be observed is non-existent
II. human behavior can only be explained through the scientific principles of cause and effect
III. mental images and ideas are only faint copies of physical sensations
IV. thinking makes it so
V. observation is the only test of truth

The CORRECT combination is:
A. All of the above B. I,II,III,V C. I,II,III,IV
D. I,III,V E. II,III,V

15. Determinism is a theory that states: 15. ___
I. Human efforts can change the world
II. Every human action can be traced to a physical cause
III. Your thoughts about what you do have no effect on your actions
IV. There is no such thing as free will
V. People are basicallv egoistic

The CORRECT combination is:
A. I,III,V B. I,II,III C. II,III,V D. I,III,IV
E. II,III,IV

16. Max Weber, an early sociologist, is well-known for 16. ___
I. using modern methodologies in his field
II. showing the connection between Puritan Protestanism and capitalism
III. developing statistical techniques in social science
IV. including values in his research and findings
V. his theories about political power

The CORRECT combination is:
A. I,II,III B. II,III,IV C. I,II,IV D. III,IV,V
E. I,III,V

17. A humanist is someone who 17. ___
I. disbelieves in God
II. promotes ethical values as the most important
III. emphasizes the power of humans to do good
IV. encourages active participation in life
V. takes advantage of findings of modern social sciences

The CORRECT combination is:
A. I,II B. III,IV C. I,III,V D. I,II,III,IV
E. II,III,IV,V

18. A necessary condition is one that 18. ___
I. helps reach a conclusion
II. must obtain for the result
III. is enough by itself to bring about a result
IV. never found for most situations in life
V. cannot be omitted without losing the result

The CORRECT combination is:
A. I,II B. II,III C. III,IV D. III,V E. II,V

19. A sufficient condition is one that 19. ___
    A. is enough by itself to bring about a result
    B. needs other conditions to satisfy all requirements
    C. is rarely found in most circumstances
    D. is optional for a given conclusion
    E. is usually equivalent for that which it is a condition of

20. Descriptive statements differ from normative ones in that the 20. ___
    I. *former* include value judgments while the latter do not
    II. *latter* include value judgments while the former do not
    III. *former* are co-extensive with ordinary empirical claims
    IV. *latter* use some word such as "good" or "bad"
    V. *former* are morally neutral

    The CORRECT combination is:
    A. I,V B. II,V C. III,IV D. IV,V E. I,III,V

21. A presupposition of an argument is 21. ___
    I. a logical conclusion to the premises
    II. a hidden premise
    III. an assumption underlying one or all the premises
    IV. useful to identify when analyzing the argument
    V. All of the above

    The CORRECT combination is:
    A. I,II B. II,III C. II,IV D. III,IV E. V

22. Buridan, a medieval logician, had an Ass (donkey) that was a rational agent, i.e., it sought to maximize its own self-interest. Placed equidistant between two entirely identical piles of hay, the donkey (given the above conditions) would decide to 22. ___
    A. randomly choose either of the two piles of hay
    B. flip a coin to choose which to begin eating
    C. starve to death in a paralysis of decision
    D. first go to one pile and eat, then to the next
    E. give up on rationality

23. A *universal* statement differs from a *general* statement in that the 23. ___
    I. *former* covers all instances while the latter does not
    II. *latter* covers all instances while the former does not
    III. *former* can be refuted by a single counter-example
    IV. *latter* can be refuted by a single counter-example
    V. *latter* constitutes much of science

    The CORRECT combination is:
    A. I,III B. I,V C. I,III,V D. II,III,V E. III,V

24. Fatalism is a theory that claims that 24. ___
I. God has prearranged all of our lives down to the last detail
II. everything that happens was meant to happen
III. nothing we do will change the future in any way
IV. people should not be held responsible for what happens in life
V. only scientific investigation explains reality

The CORRECT combination is:
A. I,II B. II,III C. III,IV D. II,III,IV
E. II,III,V

25. Aristotle was renowned for his philosophical methodology which was 25. ___
A. characterized by its empirical nature
B. in stark contrast to Platonic philosophy
C. frequently in harmony with common sense
D. a forerunner of the techniques of modern scientific method
E. All of the above

26. Teleological theories in ethics are ones that 26. ___
I. emphasize the nature of the act in question for morality
II. concern the consequences or outcomes of an act for morality
III. are sometimes called consequentialist theories
IV. require an in-depth knowledge of human nature
V. depend upon ordinary common sense notions about right and wrong

The CORRECT combination is:
A. II,III,V B. I,II,III C. I,III,IV D. I,IV,V
E. II,III

27. Deontological theories in ethics are ones that 27. ___
I. are not teleological
II. are not consequentialist
III. are consequentialist but not teleological
IV. are teleological but not consequentialist
V. emphasize the nature of an act for moral evaluation

The CORRECT combination is:
A. I,III B. I,V C. II,IV D. I,III,IV E. I,III,V

28. Kant's ethics relied heavily upon 28. ___
I. human capacity for reason
II. past experience of immoral and moral behavior
III. taking into account differences in people and situations
IV. making moral judgments universal
V. figuring out the consequences of actions

The CORRECT combination is:
A. I,V B. I,IV C. II,IV D. III,IV E. I,IV,V

29. The Categorical Imperative, according to Kant, stated that 29. ___

A. autonomy is an important moral prerequisite
B. everyone should be willing to have others do as they do
C. as long as a person does what is right, motives are irrelevant
D. orders should be followed categorically
E. None of the above

30. The following are TRUE of Kantian ethics: 30. ___

I. It is a deontological theory
II. It is a teleological theory
III. Its foundation lies in empirical fact
IV. Its foundation lies in a priori reasoning
V. It is deterministic

The CORRECT combination is:
A. I,III B. II,III C. II,IV D. I,IV E. I *only*

31. If some types of behavior are part of human nature, 31. ___

I. it is still understandable if some people do not exhibit it
II. it is not understandable if some people do not exhibit it
III. if some people do not exhibit it, it is not part of human nature
IV. it is universally found
V. All of the above

The CORRECT combination is:
A. V *only* B. II,III C. III,IV D. II,III,IV
E. I,II,III

32. Some *primary* elements in Kantian ethics are 32. ___

I. having a good will
II. justifying paternalistic behavior
III. preserving autonomy for the individual
IV. treating people as an end, not as a means
V. looking out for the self-interests of others

The CORRECT combination is:
A. I,IV B. II,IV C. III,IV D. I,III,IV
E. I,III,V

33. The principle of Universalization requires that 33. ___

I. no person make an exception of himself/herself in morality
II. what is required of one person in a given situation is required of any other similar person in a relevantly similar situation
III. moral rules apply equally to all
IV. no one should evaluate each situation on its own merits
V. everyone should evaluate each situation on its own merits

The CORRECT combination is:
A. I,II,III B. III,IV,V C. II,III,IV,V D. I,II,III,IV
E. All of the above

34. Concepts that have NO place in Kantian ethics are: 34. ___
I. duty II. intentions III. motivation
IV. consequences V. relativism

The CORRECT combination is:
A. I,II,III B. II,III,IV C. IV,V D. A *only*
E. C *only*

35. Concepts that are *crucial* to Kantian ethics are: 35. ___
I. duty II. intentions III. motivation
IV. consequences V. relativism

The CORRECT combination is:
A. I,II,III B. II,III,IV C. I,II D. I,III
E. II,III,IV

36. Relativism states 36. ___
I. nothing precisely
II. that everyone has different moral beliefs so that it is impossible to say for sure what is right or wrong
III. that what is right in one place is wrong in another
IV. that it is usually clear what is right and what is wrong
V. there is no such thing as right and wrong

The CORRECT combination is:
A. I,II B. I,II,III C. I,II,III,IV D. III,V
E. I,II,III,V

37. Popular versions of relativistic ethics include the following sorts of claims: 37. ___
A. "Who are you to judge?"
B. "Right and wrong depend upon the culture one is in"
C. "No one ought to judge others"
D. "It is wrong to make moral judgments"
E. All of the above

38. The following claim(s) of relativism is(are) logically fallacious: 38. ___
A. "Who are you to judge?"
B. "Right and wrong depend upon the culture one is in"
C. "No one ought to judge others"
D. "It is wrong to make moral judgments"
E. All of the above

39. Emotivism is a moral theory that states 39. ___
I. that moral judgments are based on intuition
II. that how much a person wants a certain consequence of an action determines whether it is right or not
III. that morality is simply a matter of how a person feels
IV. that moral judgements like "Stealing is wrong" are precisely like statements like "I like chocolate ice cream"
V. All of the above

The CORRECT combination is:
A. I,III B. II,III C. I,III,IV D. III,IV
E. None of the above

40. Cognitivism is the theory that states that moral judgments 40. ___
I. are emotional expressions
II. cannot be proved one way or the other
III. are capable of being true or false
IV. are provable by standard empirical evidence
V. are provable by analytical reasoning

The CORRECT combination is:
A. I,II,III B. II,III,IV C. III,IV D. III,IV,V
E. IV,V

41. John Stuart Mill upheld utilitarianism in 41. ___
I. precisely the same way his father, James Mill,did
II. precisely the same way Jeremy Bentham, its founder, did
III. a different way from James Mill
IV. a different way from Jeremy Bentham
V. an extensively revised version

The CORRECT combination is:
A. I,III,IV B. II,III,IV C. III,V D. III,IV
E. IV,V

42. Utilitarianism is a theory that states that 42. ___
I. a person's intentions must be taken into account when judging his/her action
II. a person;s intentions need not be taken into account when judging his/her action
III. actions should be useful before they are virtuous
IV. efficiency is the highest moral goal
V. utility is the sole principle of right conduct

The CORRECT combination is:
A. I,II B. II,IV C. I,II,V D. II,V E. II,III,V

43. Jeremy Bentham's *"hedonistic calculus"* prescribed that everyone should 43. ___
I. try to figure out how satisfied others will be by a particular act to determine whether it is right
II. determine for himself/herself whether an act will make him/her happy
III. evaluate intellectual pursuits differently from sensual ones
IV. total up the different utilities of everyone to find out which act is morally obligatory
V. total up the different utilities of only those involved to find out which act is morally obligatory

The CORRECT combination is:
A. II,III B. II,IV C. IV,V D. II,IV,V
E. II,V

44. According to Bentham, utility can be determined by 44. ___
I. each person using a *cardinal* scale for a number revealing his/her level of happiness
II. each person using an *ordinal* scale for determining his/her level of happiness
III. *objective* evidence concerning what is ultimately good for people
IV. *subjectively* evidence concerning what people like
V. using past experience as a reliable guide

The CORRECT combination is:
A. I,IV,V B. II,III,IV C. III,IV,V D. I,II E. IV,V

45. An action that is *prima facie* right is one that 45. ___
I. conforms with a moral rule or principle
II. violates a moral rule or principle
III. is the kind of action that harms people
IV. may be found later to be unjustified
V. is the kind of action that harms no one

The CORRECT combination is:
A. I,IV,V B. II,IV,V C. I,III,V D. I,II E. IV,V

46. Some actions are intrinsically wrong. 46. ___
I. Utilitarian theory *affirms* this
II. Utilitarian theory *denies* this
III. Deontological theory *affirms* this
IV. Deontological theory *denies* this
V. Emotivism *denies* this

The CORRECT combination is:
A. I,III,V B. II,IV,V C. II,III,V D. I,III E. III,V

47. Mill's revisions of Bentham's utiliarian included 47. ___
I. *widening* the role of "pleasure" to include non-sensual things
II. *reducing* the role of consequences in making moral decisions
III. *challenging* Bentham's dictum that "Pushpin is as good as poetry"
IV. *agreeing with* Bentham's dictum that "Pushpin is as good as poetry"
V. *claiming* that "Better a Socrates dissatisfied than a pig satisfied"

The CORRECT combination is:
A. II,IV,V B. II,III,V C. I,III,V D. I,V E. I,III

48. Mill corrected a serious flaw in utilitarian theory by adding the following phrase or statement to standard utilitarianism: 48.
   A. "for the greatest number"
   B. "Right actions maximize the happiness of the agent and those affected by the act
   C. "with good intentions"
   D. "The ultimate goal of morality is to promote the greatest good"
   E. "in God we trust"

49. Act-utilitarianism states that 49.
   I. moral rules should be made concerning typical actions people perform every day
   II. no act is right or wrong in itself
   III. the principle of utility need *not* be applied in every situation
   IV. the principle of utility needs to be applied in every situation
   V. past experience is not a reliable guide for judging actions

   The CORRECT combination is:
   A. I,IV,V B. II,IV,V C. I,III,V D. IV,V E. I,II

50. Rule-utilitarianism states that 50.
   I. moral principles should encompass typical actions that people perform frequently
   II. deciding the moral rightness of each act is inefficient and does not promote happiness
   III. sometimes an action ought to be performed, morally speaking, even though it does not promote the greatest happiness
   IV. no action should be performed if it itself does not increase the happiness of everyone
   V. past experiences are no basis for moral judgment

   The CORRECT combination is:
   A. II,III,IV B. I,II,III C. I,III,V D. II,IV,V E. I,II

KEY (CORRECT ANSWERS)

1.	E	11.	D	21.	D	31.	D	41.	D
2.	E	12.	B	22.	C	32.	D	42.	D
3.	D	13.	C	23.	C	33.	D	43.	E
4.	E	14.	B	24.	D	34.	C	44.	A
5.	E	15.	E	26.	E	35.	A	45.	A
6.	D	16.	C	26.	A	36.	E	46.	C
7.	A	17.	E	27.	B	37.	E	47.	C
8.	C	18.	E	28.	B	38.	E	48.	A
9.	B	19.	A	29.	B	39.	D	49.	B
10.	E	20.	A	30.	D	40.	D	50.	B

EXAMINATION SECTION

TEST 1

DIRECTIONS: Each question or incomplete statement is followed by several suggested answers or completions. Select the one that BEST answers the question or completes the statement. *PRINT THE LETTER OF THE CORRECT ANSWER IN THE SPACE AT THE RIGHT.*

1. According to Kant, the sum total of all affirmations is 1.___
   A. result B. validity C. thetics
   D. truth E. generalization

2. In logic, any proposition contrary to general opinion but capable of being supported by reasoning is termed 2.___
   A. syllogism B. postulation C. democratic
   D. thesis E. acceptable

3. Adolph Hitler's philosophical inventor was 3.___
   A. Nietzsche B. Marx C. Kant
   D. Schopenhauer E. Locke

4. _____ is inference from one premise of a syllogism to the disjunction of the conclusion and the negation of the other premises. 4.___
   A. Diallelon B. Didactics C. Diallelus
   D. Dialogism E. Dialogic

5. In Aristotelean logic, _____ was used to describe the art of specious reasoning. 5.___
   A. fallacious B. specious C. sophomoric
   D. illogic E. eristic

6. A relation of equivalence between two propositions in propositional or symbolic form is termed 6.___
   A. equity B. x and y C. logic
   D. x - y E. equipollence

7. That mode of being which consists in interaction with other things, but which, for Aristotle, is *matter clothed with form* is 7.___
   A. tension B. function C. progression
   D. logic E. existence

8. Any situation which is deliberately set up by an investigator with a view to verifying a theory is termed 8.___
   A. procedural B. experimental C. a priori
   D. controlled E. a posteriori

9. _____ is the elimination of all presuppositions of thought. 9.___
   A. Action B. Being
   C. Meditation D. Life
   E. Pure experience

10. The theory that purpose is present in all the events of the physical order is 10.___
A. Partheism B. Utilitarianism
C. Totality D. Functionalism
E. Finalism

11. _____, according to Kant, is an indubitable fact and he places it at the fulcrum of his entire system. 11.___
A. Freedom B. Harmony C. Thought
D. Faith E. Action

12. That one of the three terms in a syllogism which appears in both premises is the 12.___
A. middle term B. essential middle
C. universal D. two-dimensional object
E. bilateral

13. The suspension of judgments in regard to possible truths until they have been demonstrated to be either true or false is 13.___
A. cynicism B. inductive reasoning
C. atheism D. irony
E. methodic doubt

14. _____ is an arbitrary title given by Andronicus of Rhodes to a certain collection of Aristotelean writings. 14.___
A. Epistemology B. Ethics C. The Republic
D. Logic E. Metaphysics

15. Any procedure employed to attain a certain end is termed 15.___
A. object B. induction C. conduction
D. reason E. method

16. The reference of a mental state to something beyond itself is termed 16.___
A. supernatural B. theistic reference
C. religious D. naturalistic reference
E. transcendent reference

17. The term _____ refers to a theory of value which holds that value has an intrinsic worth apart from considerations of any particular point of view. 17.___
A. non-partisan B. timology C. irrelevant
D. aestheticism E. idealism

18. The general medium in which all events either appear to or do take place in succession is 18.___
A. space B. science C. time
D. relation E. sequitur

19. *Man is mortal; John is man; therefore, John is mortal* is an example of 19.___
A. syllogism B. analogy
C. logic D. reductio ad absurdum
E. irony

20. John Dewey would agree MOST with the following statement: 20.___
   A. Learning is solely an intellectual process
   B. One learns best by doing
   C. All men can be taught to learn at the same rate of speed
   D. One man's meat is another man's meat
   E. Think it out

21. "The theory of purpose, ends, goals, final causes; the opposite of Mechanism"is 21.___
   A. aestheticism B. theology C. finalism
   D. religion E. teleology

22. In logic, the general method of deduction which proceeds from the simple to the complex is termed 22.___
   A. reason B. synthesis C. evolution
   D. atomism E. mathematics

23. The systematized wholes of life views and tendencies, such as aestheticism, naturalism, and intellectualism, are termed 23.___
   A. synthesis B. chronology C. schema
   D. syntagma E. ideational

24. A movement to bring about a harmony of positions in philosophy or theology which are somewhat opposed or different is termed 24.___
   A. syncretism B. median pointillism
   C. arbitration D. compromise
   E. relativism

25. The name given by Nietzsche to what he deems a higher type of humanity is 25.___
   A. terminal evolution B. God
   C. racial purity D. superman
   E. supernatural

KEY (CORRECT ANSWERS)

1. C
2. D
3. A
4. D
5. E

6. E
7. E
8. B
9. E
10. E

11. A
12. A
13. E
14. E
15. E

16. E
17. B
18. C
19. A
20. B

21. E
22. B
23. D
24. A
25. D

TEST 2

DIRECTIONS: Each question or incomplete statement is followed by several suggested answers or completions. Select the one that BEST answers the question or completes the statement. *PRINT THE LETTER OF THE CORRECT ANSWER IN THE SPACE AT THE RIGHT.*

1. Which philosopher was a tutor of Alexander the Great? 1.___
   A. Plato B. Aristotle C. Socrates
   D. Xenophon E. Euclid

2. Which philosopher analyzed all reality as composed of units of force called monads? 2.___
   A. Spinoza B. Leibniz C. Roger Bacon
   D. Bergson E. Croce

3. To Henri Bergson, author of CREATIVE EVOLUTION, the evolutionary process was the basic fact of the universe. The creative impulse and cause behind the evolutionary process, according to Bergson, was 3.___
   A. the libido B. the monads C. Substance
   D. Noumenon E. Elan vital

4. The philosopher who did not trust artists and ejected them from his ideal state as dangerous and undesirable was 4.___
   A. Aristotle B. Plato C. Socrates
   D. Hegel E. Schopenhauer

5. The book in which the deed mentioned in the preceding question is recounted is 5.___
   A. REPUBLIC
   B. POLITICS
   C. POETICS
   D. THE WORLD AS WILL AND IDEA
   E. STATESMAN

6. Rousseau advocated that one should lead one's life according to 6.___
   A. reason B. the Bible
   C. social mores D. emotion
   E. the Napoleonic code

7. A precursor of the twentieth century theory that matter is composed of atoms was a Greek philosopher by the name of 7.___
   A. Thales B. Empedocles C. Anaxagoras
   D. Plato E. Democritus

8. The philospher who believed that blind, irrational will underlies all being and reality was 8.___
   A. Spencer B. Plato C. Bergson
   D. Schopenhauer E. Hegel

9. Rationalism is a doctrine which maintains that thought is the only clue to the nature of reality and of ethical good. 9.___
In which of the three groups below can ALL the philosophers mentioned be classified as rationalists?
   A. Hegel, Schopenhauer, Descartes
   B. Spinoza, Descartes, Hegel
   C. Spinoza, Bergson, Hegel

10. The twentieth century has witnessed a dethronement of God as the center and mover of the universe. 10.___
In which one of the three groups below can ALL of the philosophers mentioned be said to be directly in agreement with the cry, *God is dead*?
   A. Sartre, Camus, Nietzsche
   B. Nietzsche, Marcel, Sartre
   C. Camus, Marcel, Sartre

11. In which of the three groups below can ALL of the philosophers mentioned be said to be primarily theologians? 11.___
   A. Paul Tillich, Sidney Hook, Reinhold Niebuhr
   B. Reinhold Niebuhr, Jacques Maritain, Paul Tillich
   C. Morris R. Cohen, Reinhold Niebuhr, Paul Tillich

12. Monism is the theory that there is only one ultimate reality, whether spirit, matter, or neutral substance. 12.___
Which one of the following philosophers was NOT a Monist?
   A. Spinoza B. Zoroaster C. Thales
   D. Parmenides E. Bergson

13. Dualism is the theory that reality is composed of only two irreducible substances, matter and spirt, for example. 13.___
Which of the following philosophers was NOT a Dualist?
   A. Descartes B. Locke C. Pythagoras
   D. Anaximenes E. Zoroaster

14. Dialectical materialism is the view that reality is material and its mode of operation is through a constant struggle of opposites. This view emphasizes class struggle, believing that history proceeds only from economic drives. 14.___
Which of the following philosophers could NOT possibly be classified as a dialectical materialist?
   A. Marx B. Engels C. Lenin
   D. Bergson E. Plekhanov

15. Hedonism is a philosophical doctrine which supports pleasure as the highest good. 15.___
Which one of the following philosophers could NOT possibly be classified as a Hedonist?
   A. Epicurus B. Kierkegaard C. Bentham
   D. J.S. Mill E. Helvetius

16. The doctrine that one's highest goal should be withdrawal from the ordinary world and concentration upon the realms of the spirit, with severe self-discipline, is called 16.___
   A. Altruism B. Asceticism C. Atheism
   D. Communism E. Dogmatism

17. The word *Pragmatism* was coined by 17.___
   A. Santayana B. Dewey
   C. Bertrand Russell D. C.S. Peirce
   E. James

18. Hegel defined the Absolute as 18.___
   A. God B. Substance
   C. Matter D. Universal Reason
   E. Noumenon

19. Spinoza advocated that one should lead one's life according to 19.___
   A. emotion B. reason C. the Bible
   D. intuition E. social mores

20. The ultimate reality in Spinoza is called 20.___
   A. Mind B. Matter C. Substance
   D. Noumenon E. Monad

21. In Plato's REPUBLIC, the philosophers are 21.___
   A. workers B. statesmen C. soldiers
   D. athletes E. merchants

22. Which of the elements below is NOT a part of the nomenclature of Freudian psychoanalysis? 22.___
   A. Wish-fulfillment B. Ego
   C. Id D. Libido
   E. Elan vital

23. Which one of the following was NOT an attribute of St. Augustine's CITY OF GOD? 23.___
   A. Holiness B. Eremiticism C. Unity
   D. Apostolicity E. Catholicity

24. Many philosophers excelled in other related or unrelated fields.
Which philosophers among the following did studies PRELIMINARY to calculus and symbolic logic? 24.___
   A. Spinoza B. Leibniz
   C. Aristotle D. Leonardo da Vinci
   E. Erasmus

25. Which philosopher maintained himself through the trade of lens-grinding? 25.___
   A. Spinoza B. Leibniz
   C. Aristotle D. Leonardo da Vinci
   E. Erasmus

KEY (CORRECT ANSWERS)

1. B
2. B
3. E
4. B
5. A

6. D
7. E
8. D
9. B
10. A

11. B
12. B
13. D
14. D
15. B

16. B
17. D
18. D
19. B
20. C

21. B
22. E
23. A
24. B
25. A

TEST 3

DIRECTIONS: Each question or incomplete statement is followed by several suggested answers or completions. Select the one that BEST answers the question or completes the statement. *PRINT THE LETTER OF THE CORRECT ANSWER IN THE SPACE AT THE RIGHT.*

1. *Philosophy is history* is the phrase of 1.___
   A. Croce B. Hegel
   C. Schiller D. Foerster
   E. Bertrand Russell

2. Hugh of St. Victor was a believer in the school of 2.___
   A. Theism B. Voluntarism
   C. Transcendentalism D. Solipsism
   E. Subjective Idealism

3. A man much absorbed in the contrast of the *Dionysian* and *Apollonian* view of life, he felt the mass of mankind was an instrument for the service of the great, was 3.___
   A. Spinoza B. Schopenhauer C. Nietzsche
   D. Heraclitus E. Engels

4. TYPES OF ETHICAL THEORY was an outstanding work of 4.___
   A. F.E. Abbot B. Volkelt C. G.S. Morris
   D. Martineau E. Jodl

5. The encyclical letter, INSCRUTABILI DEI CONSILIO, was written by 5.___
   A. Pope Leo XIII B. St. Augustine
   C. St. Thomas Aquinas D. Pope Pius I
   E. John XXIII

6. ______, who wrote FINAL CAUSES, was a representative of the eclectic school. 6.___
   A. Spir B. von Sigwart C. Teichmuller
   D. G.H. Lewes E. Janet

7. ______, who wrote DE L'INTELLIGENCE, has been called the high priest of misanthropy. 7.___
   A. Ardigo B. Taine C. Ravaisson
   D. Lachelier E. Jevons

8. MAN'S PLACE IN NATURE was written by 8.___
   A. Huxley B. von Helmholtz C. Darwin
   D. Lotze E. Fechner

9. The phrase, *survival of the fittest*, was originated by 9.___
   A. Herbert Spencer B. Darwin
   C. Huxley D. Mansel
   E. Lassalle

10. WALDEN is the MOST important work of 10.___
A. Boole B. Bain C. Emerson
D. Thoreau E. Alcott

11. The EXAMINATION OF SIR WILLIAM HAMILTON'S PHILOSOPHY was written by 11.___
A. De Morgan B. F.T. Vischer
C. John Stuart Mill D. Max Stirner
E. Bauer

12. ESSAY CONCERNING HUMAN UNDERSTANDING was written by 12.___
A. John Locke B. Sir Isaac Newton
C. Richard Burthogge D. Pascal
E. George Fox

13. The man who coined the term *Pantheism* and who wrote CHRISTIANITY NOT MYSTERIOUS was 13.___
A. John Toland B. Pierre Bayle
C. John Sergeant D. Henry More
E. Geulincx

14. THREE DIALOGUES BETWEEN HYLAS AND PHILONOUS was written by 14.___
A. Mandeville B. Samuel Clark
C. George Berkeley D. Plato
E. John Norris

15. Leibniz's position, expressed in the sentence, *There is nothing in intellect which was not first in sense, except intellect itself*, answered 15.___
A. Collins' DISCOURSE ON FREE THINKING
B. MONADOLOGY
C. Collier's A NEW INQUIRY AFTER TRUTH
D. Locke's ESSAY
E. TINDAL

16. SPIRIT OF THE LAWS was the CHIEF work of 16.___
A. Swedenborg B. Hartley C. Montesquieu
D. Home E. Hume

17. L'HOMME MACHINE, a classic attempt at a materialistic interpretation of human life, was written by 17.___
A. Condillac B. Diderot C. Bolingbroke
D. d'Alembert E. La Mettrie

18. The philosopher who contributed articles on music and political economy to Diderot's ENCYCLOPÉDIE was 18.___
A. Helvetius B. Rousseau C. Voltaire
D. d'Holbach E. Lambert

19. LETTERS TO A GERMAN PRINCESS, which incorporated diagrams for indicating the relations of logical classes, was written by 19.___
A. Lessing B. d'Holbach C. Mendelssohn
D. Euler E. Price

20. Best known for his ORIGIN AND PROCESS OF LANGUAGE and ANCIENT METAPHYSICS is 20.___
A. Lord Monboddo
B. Priestley
C. Thomas Reid
D. Ethan Allen
E. Herder

21. INTRODUCTION TO THE PRINCIPLES OF MORALS AND LEGISLATION, an attempt to reduce human conduct to its ultimate motives, was written by 21.___
A. Jeremy Bentham
B. Jacobi
C. Cabanis
D. Dugald Stewart
E. Fichte

22. The METACRITIQUE OF THE PURISM OF PURE REASON was written by 22.___
A. de Tracey
B. John Hamann
C. J.F. Fries
D. Karl Krause
E. Paley

23. THE WORLD AS WILL AND IDEA was written by 23.___
A. Jefferson
B. Schopenhauer
C. Beneke
D. Coleridge
E. de Maistre

24. The three approximate means of escape from ultimate pessimism -- art, sympathy, and complete indifference to living -- were put forth by 24.___
A. Schopenhauer
B. von Hartmann
C. Rosmini-Serbati
D. Taine
E. Ecclesiastes

25. Positivism was conceived by 25.___
A. Kant
B. Carlyle
C. Comte
D. Strauss
E. Emerson

KEY (CORRECT ANSWERS)

1. B
2. D
3. C
4. D
5. A

6. E
7. B
8. A
9. A
10. D

11. C
12. A
13. A
14. C
15. D

16. C
17. E
18. B
19. D
20. A

21. A
22. B
23. B
24. A
25. C

TEST 4

DIRECTIONS: Each question or incomplete statement is followed by several suggested answers or completions. Select the one that BEST answers the question or completes the statement. *PRINT THE LETTER OF THE CORRECT ANSWER IN THE SPACE AT THE RIGHT.*

1. The theory that equates all knowledge with scientific knowledge and regards metaphysical problems as meaningless because these are not capable of empirical verification is called 1.___
   A. Scholasticism B. Logical Positivism
   C. Idealism D. Existentialism
   E. Authoritarianism

2. The philosophy which claims that existence precedes essence, that life is meaningless yet anguished by constant awareness of death, and that each individual must endow his life with meaning through engagement is known as 2.___
   A. Logical Positivism B. Scholasticism
   C. Pragmatism D. Materialism
   E. Existentialism

3. The religious doctrine which emphasizes the eternal predestination of every individual to damnation or to salvation is called 3.___
   A. Determinism B. Fatalism C. Meliorism
   D. Calvinism E. Pantheism

4. A proposition assumed without proof or an undemonstrable proof from which other proofs are deducible is called a(n) 4.___
   A. Axiom B. Syllogism C. Category
   D. Hypothesis E. Theorem

5. The theory that tragedy is an imitation of human action in which a good man comes to disaster because of some failure of character, and that the purpose of tragedy is to effect a catharsis of the emotions of fear and pity in the spectator was propounded by 5.___
   A. Tolstoy B. Plato C. Santayana
   D. Aristotle E. Croce

6. The book in which the theory mentioned in the preceding question was formulated is called 6.___
   A. WHAT IS ART? B. THE SENSE OF BEAUTY
   C. POETICS D. ION
   E. AESTHETICS

7. Which of the following philosophers could NOT possibly be classified as a Logical Positivist? 7.___
   A. Bertrand Russell B. Ludwig Wittgenstein
   C. Rudolph Carnap D. Jean-Paul Sartre
   E. Moritz Schlick

8. The method of inference from generals to particulars is called 8.___
   A. Syllogism B. Axiom C. Inference
   D. Induction E. Deduction

9. *That from which the change or the resting from change first begins* is called 9.___
   A. Final cause B. Efficient cause
   C. Being D. Becoming
   E. Mechanism

10. The concept in the preceding question marks the philosophy of 10.___
   A. Hegel B. Aristotle C. Plato
   D. Schopenhauer E. Dewey

11. Argument from cause to effect, based on knowledge independent of experience, is called 11.___
   A. a priori B. a posteriori C. cosmological
   D. ontological E. teleological

12. Argument from effect to cause, based on experience, is called 12.___
   A. a priori B. a posteriori C. cosmological
   D. ontological E. teleological

13. The branch of philosophy which investigates the essential and absolute nature of reality as a whole, or of the nature of being as such, is known as 13.___
   A. Ethics B. Aesthetics C. Apologetics
   D. Metaphysics E. Epistemology

14. The doctrine that life is the basic reality of which everything else is a form or manifestation is known as 14.___
   A. Voluntarism B. Utilitarianism
   C. Tropism D. Mechanism
   E. Vitalism

15. The theory of the origin or creation of the world is called 15.___
   A. cosmology B. epistemology C. contingency
   D. category E. cosmogony

16. *Cogito ergo sum, I think, therefore I am* is a famous expression attributed to 16.___
   A. Locke B. Hobbes C. Hume
   D. Descartes E. Spinoza

17. *Know Thyself* is a famous maxim of 17.___
   A. Aristotle B. Heraclitus C. Democritus
   D. Zeno E. Socrates

18. The doctrine that pleasure is the actual and also the proper motive of every choice is known as 18.___
    A. Idealism B. Hedonism C. Realism
    D. Naturalism E. Polytheism

19. The theory that the world is neither totally evil nor totally good and that there is a possibility of increasing the good is known as 19.___
    A. Materialism B. Mechanism C. Meliorism
    D. Monism E. Mysticism

20. The science of *being* as such, usually synonymous with metaphysics, is called 20.___
    A. Logic B. Ontology C. Pragmatism
    D. Positivism E. Nihilism

21. The study of the relation of signs and symbols to that for which they stand, which emphasizes, in general, the analysis of language for its meaning, is called 21.___
    A. Skepticism B. Solipsism C. Theism
    D. Semantics E. Voluntarism

22. The theory which seeks to justify the goodness of God in view of the evil in the world is known as 22.___
    A. Theism B. Theodicy
    C. Utilitarianism D. Tychism
    E. Solipsism

23. The metaphysical term used to indicate opposition to change or becoming, meaning *whatever is, in and of itself*, the most general predicate possible is 23.___
    A. Cause B. Class C. Being
    D. Substance E. Existence

24. The study of art values and principles of art criticism is called 24.___
    A. Aesthetics B. Ethics C. Semantics
    D. Phenomenology E. Metaphysics

25. The doctrine that truth is the practical efficacy of an idea, based on a theory of the mind as an instrument for problem-solving, which emphasizes meaning and truth in relation to the results or consequences of any idea, is known as 25.___
    A. Pragmatism B. Positivism C. Probabilism
    D. Personalism E. Pluralism

KEY (CORRECT ANSWERS)

1. B	6. C	11. A	16. D	21. D
2. E	7. D	12. B	17. E	22. B
3. D	8. E	13. D	18. B	23. C
4. A	9. B	14. E	19. C	24. A
5. D	10. B	15. E	20. B	25. A

TEST 5

DIRECTIONS: Each question or incomplete statement is followed by several suggested answers or completions. Select the one that BEST answers the question or completes the statement. *PRINT THE LETTER OF THE CORRECT ANSWER IN THE SPACE AT THE RIGHT.*

1. The theory that the test of the truth of a belief is in the way it works is called 1.___
   A. Rationalism B. Realism C. Skepticism
   D. Voluntarism E. Pragmatism

2. An example of Pessimism is found in 2.___
   A. Poincaré B. Malebranche C. Montague
   D. Ecclesiastes E. Dewey

3. The view that there are numerous kinds of irreducible reality is called 3.___
   A. Pragmatism B. Scientific Empiricism
   C. Relativism D. Pluralism
   E. Phenomenalism

4. As a final answer to basic questions, the statement, *I do not know*, is characteristic of 4.___
   A. Intuitionism B. Agnosticism C. Hedonism
   D. Neo-Realism E. Atheism

5. A representative of the school of Determinism is 5.___
   A. Marx B. Kant C. Santayana
   D. Mohammed E. Maimon

6. The belief that the conception of God can be defined and used in an acceptable fashion is called 6.___
   A. Monotheism B. Transcendentalism
   C. Atheism D. Theism
   E. Monism

7. The theory that there is no one absolutely true doctrine is known as 7.___
   A. Absolutism B. Ontology C. Allegory
   D. Relativism E. Altruism

8. The ethical doctrine supporting pleasure as the highest good is 8.___
   A. Empiricism B. Materialism C. Mysticism
   D. Hedonism E. Intuitionism

9. Excessive definiteness of belief is embodied in 9.___
   A. Dualism B. Dogmatism
   C. Determinism D. Critical Idealism
   E. Egoism

10. Plato is a representative of 10.___
A. Agnosticism B. Altruism C. Atheism
D. Conceptualism E. Absolutism

11. A contrast to the optimism of the absolute idealists is found in the pessimistic view of 11.___
A. Schopenhauer B. Hegel
C. Thomas Jefferson D. Thoreau
E. Carlyle

12. The author of LE PHILOSOPHE IGNORANT and CANDIDE is 12.___
A. d'Alembert B. Rousseau C. Helvetius
D. Diderot E. Voltaire

13. THE PRINCE is a work by 13.___
A. Pomponazzi B. Zwingli C. William James
D. Erasmus E. Machiavelli

14. MATHEMATICAL PRINCIPLES OF NATURAL PHILOSOPHY is the greatest work of 14.___
A. John Locke B. Sir Isaac Newton
C. David Hume D. Edmund Burke
E. Rousseau

15. *Cogito, ergo sum* is the classic phrase of 15.___
A. Galileo B. Descartes C. Pascal
D. Spinoza E. Socrates

16. The author of IMITATION OF CHRIST was 16.___
A. Thomas à Kempis B. Nicholas of Cusa
C. Leonardo da Vinci D. Erasmus
E. Martin Luther

17. SUMMA CONTRA GENTILES is acknowledged to be the greatest work of 17.___
A. St. Bonaventure B. Albert the Great
C. Thomas Aquinas D. Aristotle
E. Carlyle

18. The man whose teachings were assembled posthumously into the Koran was 18.___
A. Genghis Khan B. Aquinas
C. Joachim D. Mohammed
E. John of Damascus

19. The CONFESSIONS were written by 19.___
A. St. Thomas Aquinas B. Plutarch
C. Gregory the Great D. St. Augustine
E. Maximus the Confessor

20. The leader of the Modalists was 20.___
A. Cyprian B. Plotinus C. Origen
D. Sabellius E. Diogenes

21. THE HAPPY LIFE was written by 21.___
A. Barnabas B. Paul C. Ignatius
D. Seneca E. Plutarch

22. *The unexamined life is not to be lived* is attributed to 22.___
A. Socrates B. Plato C. Euclid
D. Aristotle E. Cratylus

23. THE CRITIQUE OF PURE REASON is associated with 23.___
A. Hegel B. Kant C. Schopenhauer
D. Hume E. Aristotle

24. A philosopher of the pre-Socratic period is 24.___
A. Thales B. Diderot C. Engels
D. Kierkegaard E. Bonaventure

25. The belief that the universe taken as a whole is God is expressed in 25.___
A. Pantheism B. Monism C. Panpsychism
D. Asceticism E. Methodism

KEY (CORRECT ANSWERS)

1. E
2. D
3. D
4. B
5. D

6. D
7. D
8. D
9. B
10. E

11. A
12. E
13. E
14. B
15. B

16. A
17. C
18. D
19. A
20. D

21. D
22. A
23. B
24. A
25. A

TEST 6

DIRECTIONS: Each question or incomplete statement is followed by several suggested answers or completions. Select the one that BEST answers the question or completes the statement. *PRINT THE LETTER OF THE CORRECT ANSWER IN THE SPACE AT THE RIGHT.*

1. According to Kant, theoretical (or scientific) reason is 1.___
   A. not limited to experience alone
   B. valid
   C. limited to *objects of possible experience*
   D. confined to experimentation
   E. the labeling of things which defy labels

2. Which of the following questions was NOT posed by Kant? 2.___
   A. How far is metaphysics possible?
   B. What may I hope?
   C. Am I here for a purpose?
   D. How is pure mathematics possible?
   E. What ought I do?

3. Kant believed that morality consists of actions in accordance with principles, and that humanity 3.___
   A. should dispense with the egotistical belief that it is all-important
   B. must be regarded as the final end
   C. must be freed from restrictive bonds of principles that give material success
   D. must not be regarded as the final end
   E. must conform to logical principles

4. Since *Duty* is a key word in the ideas of Kant, he would be MOST likely to reject 4.___
   A. Voluntarism
   B. Criticism
   C. Hedonism
   D. Critical Idealism
   E. Asceticism

5. Leibniz propounded a new system of philosophy in which he maintained that the ultimate elements of the universe are 5.___
   A. compressed into space-pockets released only by impact
   B. individual centers of force
   C. from one supreme force
   D. manipulated by God alone
   E. born of one infinite substance

6. His skepticism about *necessary connection* has remained a stimulus to philosophic controversies. His significant contribution to philosophy was his classic questioning of many nationalistic assumptions. 6.___
   We are speaking of
   A. Royce B. Mill C. Fries D. Maimon E. Hume

7. When asked what was Easy, Thaler said 7.___
   A. Believing
   B. To advise another
   C. To be truthful to oneself
   D. To define the Truth
   E. To know one's self

8. Those who believe that salvation comes through possession of superhuman knowledge are called 8.___
   A. Secularists
   B. Salvationists
   C. Idealists
   D. Gnostics
   E. Animists

9. The teachings of Plotinus differ from those of Philos in that 9.___
   A. they did not emphasize Jewish tradition
   B. Philo did not believe in God as the Ultimate Reality
   C. Plotinus did not believe in God as the Ultimate Reality
   D. Philo did not emphasize Jewish traditions
   E. Plotinus believed in Neo-Platonism

10. William of Auvergne opposed the _____ doctrine of the eternity of the world. 10.___
   A. Indian
   B. Arabic
   C. Hebrew
   D. Hellenistic
   E. Socratic

11. Albert the Great followed Augustine's 11.___
   A. doctrines on truth
   B. doctrines on humanity
   C. belief regarding the *dual system*
   D. belief that the Milky Way is made up of many stars
   E. theory of the creation of the world out of nothing

12. Thomas Aquinas had in common with Aristotle the belief that 12.___
   A. all life is movement
   B. all motion is preconceived action
   C. the mind is the *form* of the body
   D. the soul is the *form* of the body
   E. the mind is the principle of life

13. Carnap's work has been devoted ESPECIALLY to 13.___
   A. Logical Empiricism
   B. Spiritualism
   C. Formal Logic
   D. Mysticism
   E. Dualism

14. Manegold believed that theology 14.___
   A. is not subject to logic
   B. has its roots in the minds of idealists
   C. is the study of superstitions
   D. is only for the minds of fools
   E. is worthless to a philosopher

15. According to Aristotle, everything is either _____ or form. 15.___
A. air B. substance C. matter
D. nature E. liquid

16. The oldest Indian philosophic system is 16.___
A. Peyoteism B. Confucianism C. Yogaism
D. Sankhya E. Yajnavalkya

17. A definition of the term *Buddha* is 17.___
A. one who believes that *God and man dwell in the same temple*
B. one who broods in the temple
C. *Enlightened One who enlightens*
D. a man who has found *spiritual identity*
E. a God who has sympathy for all

18. Few doubt that Spinoza was indebted to Crescas for his 18.___
A. understanding of the Jewish doctrines of freedom
B. conversion to Catholicism
C. concept of the universe
D. concept of numbers
E. concept of nature

19. The statement, *There are no other worlds than the one system in which the earth is situated*, would be adamantly REJECTED by 19.___
A. Cusanus B. Aristotle C. Cleanthes
D. Crescas E. Bruno

20. Crescas _____ Aristotle's physics, metaphysics, and axiology. 20.___
A. accepted B. upheld C. rejected
D. believed E. evolved

21. Saadia taught that faith is not opposed to reason but 21.___
A. only to the idea of reason
B. is granted by God only
C. only to pseudo-reason
D. reason is opposed to belief
E. only what reason stands for

22. Saadia criticized Plato's _____ and refuted gnostic doctrines. 22.___
A. concept of nature B. technical terms
C. cosmology D. religion
E. idealism

23. In Aristotle, the essence, or entelechy, of a natural or organic body is 23.___
A. the Will B. the Heart C. the Mind
D. Vision E. the Soul

24. Polytheism is the OPPOSITE of 24.___
- A. many Gods who rule the mind
- B. Monotheism
- C. contemplation of the possibilities of *ideas* being Gods
- D. spiritual belief of any kind
- E. *spiritual* ideas over intelligence

25. Hobbes believed that the MOST effective form of government was 25.___
- A. an absolute monarchy
- B. anarchy
- C. democracy
- D. a triumvirate
- E. a republic

KEY (CORRECT ANSWERS)

1. C
2. C
3. B
4. C
5. B
6. E
7. B
8. D
9. A
10. B
11. E
12. D
13. C
14. A
15. C
16. E
17. C
18. C
19. B
20. C
21. C
22. C
23. E
24. B
25. A

EXAMINATION SECTION

TEST 1

DIRECTIONS: Each question or incomplete statement is followed by several suggested answers or completions. Select the one that BEST answers the question or completes the statement. *PRINT THE LETTER OF THE CORRECT ANSWER IN THE SPACE AT THE RIGHT.*

1. In the developing field of human gene therapy, which is generally considered the MOST ethically sound end for treatment? 1.___
   A. Therapeutic repair
   B. Alternative creation
   C. Composite trait selection
   D. Subjective enhancement

2. Which of the following was NOT a belief embraced by the American counter-cultural movement of the 1960s? 2.___
   A. The placement of sensory experience ahead of conceptual knowledge
   B. Existential emphasis on being rather than doing
   C. The rejection of mastery over nature
   D. Valuing detachment, objectivity, and noninvolvement as methods for finding truth

3. According to inquiries of American workers, APPROXIMATELY what percentage of people claim to be currently performing at the level that demonstrates the full range of their capabilities? 3.___
   A. 10 B. 25 C. 50 D. 75

4. Currently, which is NOT known to be an element of the most effective welfare/entitlement programs in the United States? 4.___
   A. Small offices with close ties to communities and local employers
   B. A flat bureaucratic hierarchy with a limited number of levels
   C. Cautious, custodial leadership with fixed policies and procedures
   D. Little political input from local or state politicians

5. MOST professional codes of ethics 5.___
   A. embody a broad picture of expected moral character
   B. are voluntary
   C. provide specific guidance for action in situations
   D. are decided by objective ethicists outside of the profession

6. Currently, the MOST important factor determining a child's success in an American school is the 6.___
   A. student's racial/ethnic background
   B. student's financial standing
   C. education level of the student's parents
   D. student's sense of responsibility for his/her performance

7. Since the 1920s, the research of advertising companies has revealed that the most effective ads generally produce what sort of response from potential consumers? 7.___
   A. The belief that the product or service is of high quality
   B. A strong emotional reaction
   C. The belief that the product or service is economical
   D. A desire for time-saving convenience

8. Of the following, the LEAST important determining factor for an individual's personal ethic is 8.___
   A. reason B. religion C. emotion D. law

9. Each of the factors below has contributed to the change in American political campaign ethics EXCEPT 9.___
   A. party reforms devaluing individual voters
   B. technological innovations
   C. lengthening campaigns
   D. shifting demographics

10. Contemporary broadcasts in American television are designed to promote each of the following conditions among viewers EXCEPT 10.___
    A. awareness B. understanding response
    C. interest D. knowledge

11. To early American Calvinists, the PRIMARY spiritual/religious purpose of work was to 11.___
    A. secure a place with God in the afterlife
    B. produce material comforts
    C. avoid vacant time that might inspire evil thoughts
    D. to assure onself that he/she was a member of the elected inhabitants of God's kingdom

12. In questioning the moral rules that apply in any given situation, an individual is operating on the _____ level in the determination of what is good or right. 12.___
    A. expressive-evocative B. moral
    C. ethical D. post-ethical

13. Which of the four guiding principles of modern medicine specifically resulted from the knowledge of experiments that were performed on unknowing subjects? 13.___
    A. Beneficence B. Nonmaleficence
    C. Justice D. Autonomy

14. The majority of American citizens currently view employment in terms of all of the following EXCEPT 14.___
A. status
B. expression of self-interest
C. social obligation
D. economic necessity

15. Which of the following is LEAST likely, in the current American work environment, to improve standards for American workers? 15.___
A. Making employees personally accountable for job performance
B. Assigning workers a role in planning tasks
C. Division of labor into more specialized roles
D. Direct feedback from consumers to workers

16. According to most contemporary opinion polls, which of the following characteristics is considered to be LEAST important to the effectiveness of a public official? 16.___
A. Prudence
B. Practical wisdom
C. Honesty
D. Courage

17. Generally, the current social trends in American demographics show that most people value employment primarily for each of the following reasons EXCEPT 17.___
A. economic benefits
B. avoiding the embarrassment of public assistance
C. the inherent morality in work and labor
D. peripheral benefits such as health care and insurance

18. The contemporary revolution in the communication media, especially broadcast, is generally thought to be a product of 18.___
A. the way in which most American lives are lived
B. the way most Americans believe their lives should be lived
C. fictions of life that are contrived by a limited corps of media personnel
D. nostalgic reflections on life as it used to be

19. The pattern of distribution reflecting the movement of the American personal work ethic throughout the country's early population is from the 19.___
A. lower class upward
B. upper class downward
C. middle class to the lower class
D. middle class into both the upper and lower classes

20. Generally, in the latter part of the twentieth century, M.B.A.-trained managers increased short-term profits in large corporations (such as steel and automobile industries) through each of the following practices EXCEPT 20.___
A. denial of funds for research and product development
B. sacrificing product quality

C. elimination of funds for modernization of equipment and facilities
D. widespread cuts in labor wages

21. According to the utilitarian school of ethics, the MOST reliable source of moral standards in any society is 21.___
A. the laws formulated by a democratically-elected government
B. the discourse of an elite corps of intellectual moralists
C. the opinions and convictions of the majority
D. a randomly chosen individual citizen

22. Generally, the field of bioethics began with the _____ principle as its primary concern. 22.___
A. quality of life
B. cost-efficiency of individual treatment
C. collective good of individual treatment
D. sanctity of life

23. After World War I, which of the following conditions did NOT contribute to a change in the American work ethic? 23.___
A. Widespread desire for emotional release
B. Material prosperity
C. A slackening sense of religious obligation
D. Production of nonessential *leisure* products such as radios and automobiles

24. The GREATEST source of income for most members of the House and Senate is from 24.___
A. government salary
B. outside jobs
C. unearned income from dividends and investments
D. honoraria for speaking engagements

25. In general, the effect of American television news broadcasts in recent years has shifted citizen involvement in government from 25.___
A. response to initiation
B. detachment to involvement
C. crisis to policy-making
D. action to reaction

KEY (CORRECT ANSWERS)

1. A	6. D	11. D	16. C	21. C
2. D	7. B	12. C	17. C	22. D
3. B	8. D	13. D	18. C	23. C
4. C	9. A	14. C	19. D	24. C
5. A	10. B	15. C	20. D	25. D

TEST 2

DIRECTIONS: Each question or incomplete statement is followed by several suggested answers or completions. Select the one that BEST answers the question or completes the statement. *PRINT THE LETTER OF THE CORRECT ANSWER IN THE SPACE AT THE RIGHT.*

1. Which of the following was NOT an element that contributed directly to the growth of the work ethic in America's early agricultural society? 1.___
   A. Abundant, easily-tapped natural resources
   B. A free-enterprise economy
   C. Immediate, tangible rewards for labor
   D. Religious freedom

2. The growth of American manufacturing companies generally resulted in each of the following conditions EXCEPT 2.___
   A. swelling of corporate bureaucracies
   B. decreased pressure on management personnel
   C. expansion of work responsibilities for laborers
   D. discouragement of competition from developing companies

3. Which of the following duties is NOT considered to be an appropriate use of personal staff by Congressional members? 3.___
   A. Coordinating meetings
   B. Answering correspondence
   C. Writing speeches and correspondence
   D. Distributing campaign literature

4. The trend in current American education, especially at the university level, teaches ethics in a social context that promotes 4.___
   A. the development of individual virtue
   B. ethical relativism
   C. ethics are most important in the professional arena
   D. public policy as the primary agent of principles

5. Currently, the function of the American health care industry that would serve the population MOST effectively and inexpensively is perceived to be 5.___
   A. expansion of emergency services
   B. increased specialization
   C. prevention of health crises
   D. curing afflicted individuals

6. In recent years, the majority of non-partisan social policy analysts attribute the persistence of poverty in the United States to 6.___
   A. social barriers
   B. economic injustice

C. systematic discrimination
D. lack of attractive work incentives

7. Which of the following is NOT viewed by many physicians and bioethicists as an inevitable result of the unfolding knowledge of the human genome? 7.___
A. The patenting of human genes
B. An explosion in legal costs
C. An increased abortion rate
D. The proliferation of new genetic diseases

8. Which is NOT considered to be one of the primary ethical standards against which public behavior is to be judged in a democratic society? 8.___
A. Rationality
B. Personal convictions
C. Democratic process
D. Choice

9. Generally, the legal precedents in the United States have been MOST likely to award decision-making rights concerning a child to the _____ mother. 9.___
A. genetic
B. gestational
C. social
D. spiritual

10. During which decade did the installation of computer-programmed machinery begin to replace the human skill requirement in many industrial jobs? 10.___
A. 1960s B. 1970s C. 1980s D. 1990s

11. Which of the following is NOT a factor contributing to the Federal government's inability to set behavior standards for the recipients of its entitlement programs? 11.___
A. Limitations of Federal constitutional authority
B. Implementation difficulties
C. The differing and conflicting approaches of policy analysts
D. Lack of capital to transform current programs

12. In which American service industry are ethical codes considered to be voluntary? 12.___
A. Legal representation
B. Health care
C. Financial/tax services
D. Radio and television broadcasting

13. Which ethical code would MOST likely condemn the distribution of disinfected hypodermic syringes to known drug addicts in order to prevent the spread of disease? 13.___
A. Teleological
B. Deontological
C. Utilitarian
D. Existential

14. In contemporary American politics, the power of a campaign is rooted in 14.___
A. facts
B. issues
C. causes and effects
D. stories and rites

15. As they were originally designed, affirmative action practices in America were intended to promote 15.___
   A. the establishment of racial quotas in hiring practices
   B. active recruitment of qualified applicants from historically excluded groups
   C. lowering of hiring and acceptance standards among certain groups
   D. a workforce and student population that was proportionally matched to the population at large

16. The American work ethic received its GREATEST blow among educated professionals (professors, physicians, therapists) from 16.___
   A. the human potential movement in American psychology
   B. the counter-cultural movement among American students
   C. lack of pride in work results
   D. the entertainment industry

17. The term for the shared values of a political democracy is 17.___
   A. over-ethic
   B. public interest
   C. public opinion
   D. policy

18. In general, the use of television as a medium for campaign broadcasting in the United States is considered to have resulted in each of the following EXCEPT 18.___
   A. decreased voter turnout
   B. longer speeches by candidates
   C. increased cynicism
   D. lower levels of political knowledge

19. In a democratic society, the _____ is usually considered to be the agent of ethical conduct. 19.___
   A. individual citizen
   B. elected official
   C. academic scholar
   D. information media

20. In the last few decades, personal staff work forces hired by House and Senate members have 20.___
   A. decreased by about half
   B. remained about the same size
   C. doubled
   D. tripled

21. Which body of ethical thought has its roots MOST firmly in American traditions? 21.___
   A. Phenomenology
   B. Utilitarianism
   C. Pragmatism
   D. Existentialism

22. Most ethical codes for physicians hold each of the following conditions to be signs of incompetence for informed consent in patients EXCEPT the inability to 22.___
   A. offer risk/benefit reasons
   B. communicate or articulate specific reasons
   C. grasp the situation
   D. reach a reasonable decision

23. Among typical American low-wage workers, the GREATEST sense of personal shame results from 23.___
    A. failure to meet social obligations
    B. poor performance on the job
    C. low status of individual position
    D. poor living conditions

24. Which ethical code is based on the possible good produced by behaviors? 24.___
    A. Deontological
    B. Epistemological
    C. Teleological
    D. Existential

25. The PRIMARY purpose for sperm banks in the United States is shifting toward 25.___
    A. a private industry for trait selection
    B. public storage to be accessed by women who want to bear children
    C. donation for research purposes
    D. storage for men who want to insure their own ability to father children in the future

---

KEY (CORRECT ANSWERS)

1. D
2. C
3. D
4. D
5. C

6. D
7. C
8. B
9. B
10. C

11. D
12. D
13. B
14. D
15. B

16. A
17. B
18. B
19. A
20. D

21. C
22. B
23. C
24. C
25. A

TEST 3

DIRECTIONS: Each question or incomplete statement is followed by several suggested answers or completions. Select the one that BEST answers the question or completes the statement. *PRINT THE LETTER OF THE CORRECT ANSWER IN THE SPACE AT THE RIGHT.*

1. The trend in the American health care industry is seen to have shifted its emphasis toward 1.___
   A. research
   B. therapeutic intervention
   C. prevention of disease
   D. curing afflicted individuals

2. In which type of government are ethical concerns MOST prominently at issue? 2.___
   A. Monarchy
   B. Totalitarian dictatorship
   C. Democracy
   D. Collectivist/communist

3. The professional campaigners who currently dominate the American political scene view the American public as capable of responsibly judging all but ______ in a candidate. 3.___
   A. charisma
   B. image
   C. reliability in performing public duties
   D. current stance on specific issues

4. What was the initial, primary purpose for sperm banks in the United States? 4.___
   A. A private industry for trait selection
   B. Public storage to be accessed by women who want to bear children
   C. Donation for research purposes
   D. Storage for men who wanted to insure their own ability to father children in the future

5. Which of the following conditions was NOT responsible for the severe damage inflicted upon the American work ethic in the mid-1980s? 5.___
   A. Fractionation of jobs
   B. Intense skill requirements in manufacturing jobs
   C. Dominant values of pleasure and entertainment
   D. Emphasis on short-term profits

6. In mainstream television news broadcasts in the United States, video footage is increasingly used PRIMARILY to 6.___
   A. illustrate a news story
   B. slow the pace of a broadcast
   C. substitute for narrative and analysis
   D. provide facts and statistics

7. Generally, in the late twentieth century, the trend in American social policy shifted responsibility toward which element of society? 7.___
   A. Family B. Government
   C. Smaller communities D. The individual

8. During which decade did alarm from corporate managers about the threat of foreign competition FIRST become widespread? 8.___
   A. 1950s B. 1960s C. 1970s D. 1980s

9. Which characteristic is an example of the *golden mean* advocated by traditional codes of ethics? 9.___
   A. Humility B. Modesty
   C. Pride D. Arrogance

10. In American politics, which of the following factors is NOT a reason for the population's difficulty in judging a candidate's ethical competency? 10.___
   A. Length of campaigns making occasions for assessment highly variable
   B. Availability of printed media
   C. Most candidates and campaigners viewing ethical standards as situational
   D. Increasingly shorter speeches and interviews

11. The field of bioethics has gradually shifted toward the _____ principle as its PRIMARY concern. 11.___
   A. quality of life
   B. cost-efficiency of individual treatment
   C. collective good of individual treatment
   D. sanctity of life

12. Beginning in the early twentieth century, which pattern of activity was considered MOST desirable to American workers? 12.___
   A. Sustained work interrupted by brief periods of vigorous recreation
   B. Long periods of leisure interrupted by necessary work tasks
   C. Sustained work interrupted by brief periods of idle resting time
   D. Sustained religious observance interrupted by brief periods of intense labor

13. The distribution of condoms among high school students in order to prevent the spread of disease would be MOST effectively justified by which ethical code? 13.___
   A. Deontological B. Teleological
   C. Phenomenological D. Existential

14. In contemporary American society, the vast majority of abortions are the result of indications from _____ concerns. 14.___
   A. sociological B. maternal
   C. fetal D. humanitarian

15. Beginning in the 1920s, the American advertising industry found that consumers were influenced PRIMARILY by appeals to 15.___
A. vanity B. sex drive C. reason D. hunger

16. Which of the following is a belief embraced by the American counter-cultural movement of the 1960s? 16.___
A. Rejection of rules that interfere with natural function and expression
B. Devaluing the search for truth through experience and participation
C. Adherence to organization and rationalization
D. Emphasis on the elements of nature illuminated by science, rather than the unknown or mystical

17. Which of the following is NOT a formal perquisite granted to Congressional members? 17.___
A. Franking privileges
B. Discounts from Capitol shops and services
C. Immunity from certain legal actions
D. Allowances for travel and communication

18. Currently, _____ is known to be an element of the most effective welfare/entitlement programs in the United States. 18.___
A. large, centrally-located offices
B. frequent political input from politicans and community leaders
C. an informal management style, with no rigid rules for conduct
D. a rigidly-controlled bureacracy, dominated by management personnel

19. The type of gene therapy that affects the subject's entire lineage from the point of intervention is 19.___
A. mitotic intervention B. germ-cell therapy
C. zygote therapy D. T-cell therapy

20. In general, the established evaluations of contemporary American service-industry employees (such as operators, agents, and clerks) measure the efficiency of each worker according to the 20.___
A. speed with which each customer is handled
B. quality of services rendered
C. customer satisfaction
D. profit-per-transaction

21. Which issues lies at the heart of modern Western ethics? 21.___
A. Fate B. Guilt
C. Meaninglessness D. Justice

22. Among typical Japanese low-wage workers, the GREATEST sense of personal shame results from 22.___
A. failure to meet social obligations
B. poor performance on the job
C. low status of individual position
D. poor living conditions

23. Which of the following is NOT generally considered to be a value or condition that would contribute to the restoration of the United States to economic vitality? 23.___
A. Collectivism
B. Individualism
C. Patriotism
D. Materialism

24. In the United States' mainstream television news broadcasts, the top stories for any given report are increasingly decided by 24.___
A. the perceived impact of the story on the public
B. current trends in the social and political fabric
C. pressure from corporate sponsors
D. the availability and interest level of video footage

25. In the developing field of human gene therapy, which is generally considered the MOST ethically unsound end for treatment? 25.___
A. Therapeutic repair
B. Experimental modification
C. Composite trait selection
D. Subjective enhancement

KEY (CORRECT ANSWERS)

1. D
2. C
3. D
4. D
5. B
6. C
7. B
8. C
9. B
10. B
11. A
12. B
13. B
14. A
15. B
16. A
17. B
18. C
19. B
20. A
21. C
22. B
23. A
24. D
25. D

TEST 4

DIRECTIONS: Each question or incomplete statement is followed by several suggested answers or completions. Select the one that BEST answers the question or completes the statement. *PRINT THE LETTER OF THE CORRECT ANSWER IN THE SPACE AT THE RIGHT.*

1. In America's formative years, which of the following was considered to be an enemy to the personal work ethic? 1.___
   A. Individualism
   B. Periods of recreation
   C. Idle time between work tasks
   D. Religious observances that interrupted work

2. Which of the following does NOT generally describe the administration of contemporary American political campaigns? 2.___
   A. Rhetorical B. Localized
   C. Corporate D. Mass-mediated

3. In contemporary American society, the extent of an individual's right to health care is increasingly decided by the individual's 3.___
   A. geographical proximity to the appropriate specialist
   B. severity of need
   C. awareness of the need situations
   D. ability to pay for treatment

4. Which news category has shown the GREATEST coverage increase in the last decade among mainstream news media in the United States? 4.___
   A. Politics
   B. Entertainment
   C. Social trends and problems
   D. Science and the environment

5. Since the 1920s, _____ workers have been LEAST likely to complain about the unpleasantness of working conditions in America. 5.___
   A. unskilled B. farmers and agricultural
   C. white-collar office D. low-level service

6. Which ethical code is based on personal duty? 6.___
   A. Deontological B. Epistemological
   C. Teleological D. Existential

7. The term for the element of a physician's ethical code that permits him/her to withhold information from a patient who is too frightened, unstable, or depressed to deal effectively with the information is 7.___
   A. benign malpractice B. therapeutic privilege
   C. nonmaleficence D. judicious restraint

8. Social policy in the United States, especially at the federal level, is currently thought by analysts to be oriented toward individual 8.___
   A. abilities B. obligations
   C. claims D. potentials

9. The characteristic MOST likely to restore the work ethic among industrial and service workers is 9.___
   A. obedience B. specialization
   C. autonomy D. detachment

10. Due to the nature of contemporary political campaigns, American voters are LEAST likely to receive impressions about a candidate's potential in the area of 10.___
   A. charisma B. issues orientation
   C. personal integrity D. competence

11. Each of the following is considered to be a component of America's original personal work ethic EXCEPT 11.___
   A. the gradual accumulation of material comforts
   B. a perceived moral duty to work hard
   C. pride in work skills and the quality of the resulting product or service
   D. a willingness to exert high effort to achieve material well-being

12. After their initial refusal to address the production of cheaper, better automobiles by foreign competitors, the Big Three responded by investing billions of dollars in the late 70's in 12.___
   A. robotics and machines that would eliminate paid factory positions
   B. product research and development
   C. quality-control programs
   D. personnel training and education

13. In America, the modern media GENERALLY views the consumer population as a 13.___
   A. set of differentiated groups
   B. single society
   C. confederation of individuals
   D. bureau of a few limited classes

14. Legally, a frozen human embryo's status is currently that of 14.___
   A. an individual with the right of self-determination
   B. an individual who is, until implantation, the property of an appointed legal guardian
   C. the property of the mother, who will decide its fate
   D. joint property requiring consent of both parents for development/implantation

15. Which pattern of activity was considered MOST desirable to workers in the early American agricultural society? 15.___
   A. Sustained work interrupted by brief periods of vigorous recreation
   B. Sustained work interrupted by brief periods of idle rest
   C. Intermittent periods of work with generous intervals of idle rest
   D. Sustained religious observance interrupted by brief periods of intense labor

16. The Industrial Revolution was partly responsible for the erosion of the work ethic in America because it 16.___
   A. lowered the standard of living for workers and consumers
   B. accelerated production and created more leisure time for workers
   C. reduced the workers' pride through repetitive, assembly-line tasks
   D. made managers and executives less likely to push workers to excel

17. Generally, what is the MOST influential source of values for those Americans who still practice the work ethic? 17.___
   A. Family
   B. School
   C. Work environment
   D. Religious beliefs

18. Most ethicists believe a person's ethical character to be PRIMARILY formed by 18.___
   A. societal rule and law
   B. conscious actions
   C. religious affiliation
   D. study of moral rules

19. Which of the following is NOT a category of concern currently thought to contribute to nonwork among American individuals? 19.___
   A. Economic
   B. Political
   C. Cultural
   D. Religious

20. In general, during the period from 1950-1980, how did the television entertainment industry influence the American work ethic? 20.___
It
   A. poked fun at working-class families and lifestyles
   B. glorified the virtues of hard work
   C. avoided any reference or allusion to the middle class
   D. dealt extensively with social issues and the American workplace

21. Which word BEST describes the contemporary standards by which the population currently makes ethical assessments of public figures? 21.___
   A. Situational
   B. Traditional
   C. Absolute
   D. Relative

22. The traditional definition of equality in America includes all but _____ characteristics of the individual. 22.___
    A. rights
    B. needs
    C. obligations
    D. freedoms

23. Most ethical codes for physicians hold each of the following conditions in a patient to be signs of incompetence for informed consent EXCEPT the inability to 23.___
    A. offer a preference
    B. provide a rational reason
    C. pay for expensive, specialized services
    D. understand information

24. Which of the following topics is LEAST likely to be debated today, given the current state of American politics? 24.___
    A. Individual freedoms
    B. The extent of federal government
    C. Individual responsibilities
    D. Individual needs

25. With the use of their franking privileges, Congressional members are allowed to distribute each of the following types of postal material EXCEPT 25.___
    A. public opinion poll results
    B. biographies and pictures
    C. holiday greetings
    D. press releases that deal with legislative activity

KEY (CORRECT ANSWERS)

1. C
2. B
3. D
4. B
5. B

6. A
7. B
8. C
9. C
10. B

11. A
12. A
13. A
14. D
15. A

16. C
17. A
18. B
19. D
20. A

21. A
22. C
23. C
24. C
25. C

EXAMINATION SECTION

TEST 1

DIRECTIONS: Each question or incomplete statement is followed by several suggested answers or completions. Select the one that BEST answers the question or completes the statement. *PRINT THE LETTER OF THE CORRECT ANSWER IN THE SPACE AT THE RIGHT.*

1. In 1980, what Vietnam-era behavior by General Westmoreland and other top-ranking military officials was called into question by the news media? 1.___
   A. The bombing of Cambodia
   B. Human rights abuses in American prisoner-of-war camps
   C. Ordering military attacks on civilian population centers
   D. Artificial inflation of enemy numbers to influence political support of the war effort

2. Soon after the state of Oregon ended Medicaid funding for transplants, a young child named Coby Howard died of leukemia because his parents could not raise enough money for a transplant. 2.___
   The federal court held that a state is
   A. liable for civil wrongful-death suits if it suspends funding for life-saving operations
   B. obligated to provide funding, up to a certain amount, for live-saving operations
   C. liable for criminal negligence charges if it suspends funding for life-saving operations
   D. not obligated to provide funding for transplants, despite the likelihood of death for patients who go untreated

3. Truck driver Reginald Denny, a beating victim during the Los Angeles riots of 1992, eventually filed a civil lawsuit against 3.___
   A. the municipal authorities for failing to control the riotous activity that encouraged his assault
   B. his attackers for injuries sustained during the assault
   C. the National Guard for inept handling of the riot
   D. his employer for sending him into an area that was dangerous for whites to enter

4. According to public opinion, what was former President Nixon's GREATEST crime in association with the Watergate affair? 4.___
   A. The initial burglary at the Democratic headquarters
   B. The subsequent cover-up of Executive involvement
   C. Spying on White House conversations through secret bugging devices
   D. Blackmailing key figures who were potentially incriminating

5. Currently, what is the APPROXIMATE percentage of American workers who have a *just cause* provision for termination included in their employment contracts? 5.___
A. 30% B. 50% C. 75% D. 95%

6. Which of the following behaviors is NOT expressly forbidden by the American government's official code of ethics? 6.___
A. Using one's office or position for private financial gain
B. Withholding information for private financial gain
C. Soliciting or receiving money for advice or assistance
D. Offering or receiving gifts from constituents

7. Currently, what percentage of Americans support therapeutic abortions (abortions that are administered due to a threat to the mother's health)? 7.___
A. 10% B. 25% C. 50% D. 90%

8. In order to break the Watergate story for the Washington POST, journalists Carl Bernstein and Bob Woodward lied to some of the figures involved in order to elicit incriminating responses. 8.___
Which of the following elements of the journalist's informal code of ethics was violated in this case?
A. Maintaining balance B. Avoiding harm
C. Maintenance of purpose D. Avoiding bias

9. Which of the following is NOT considered by a majority of Americans to be a valid ethical argument against adultery? 9.___
A. It is conducted regardless of a person's feelings of love.
B. It usually involves deception.
C. Its nature inevitably causes the breaking of a promise.
D. It is harmful to the institution of marriage.

10. The 1991 case of Helga Wanglie concerned a comatose woman who had previously expressed a wish to be aggressively treated regardless of circumstances. Despite the hospital's attempt to terminate what they considered to be pointless treatment, a court ordered that Mrs. Wanglie should remain on a ventilator indefinitely. 10.___
The legal rationale for this decision was that
A. an individual's preferred mode of treatment is an unconditional legal right
B. life-sustaining treatment cannot be denied even brain-dead patients under any circumstances
C. an individual's will in such cases was contingent upon the informed consent of immediate family members
D. an individual cannot be denied life-sustaining treatment against his/her will

11. The U2 incident of the Eisenhower administration, in which an American spy plane was shot down while in Soviet airspace, was met by an American Executive denial that the plane was meant for anything other than *reconnaissance*. Later, the government was forced to admit it was, in fact, a spy mission. 11.____
The general effect on the American public was
   A. an escalation of anti-communist sentiment
   B. weakening faith in United States leadership
   C. acceptance of the American response
   D. a call for an extensive investigation into the incident

12. An increasing trend among radical liberals in America is 12.____
   A. campaigns of violent protest and dissent
   B. a movement away from the political process toward education and social reform
   C. the adoption of methods and an exclusive moralism similar to the religious right
   D. a movement toward moderation in goals and outcomes

13. Which of the following is NOT a problem with the way in which political campaigns are currently being conducted in the United States? 13.____
   A. Corporations or individuals with large cash reserves may drown out the message of the majority
   B. Many campaigns are conducted at a personal, one-on-one level
   C. Lack of laws and regulations concerning the conduct of campaigns
   D. The public does not have access to enough information to make an informed decision

14. Mary Beth Whitehead, a surrogate mother who legally contracted to bear a child for a married couple and later decided to sue for custody of the child, was denied custody because 14.____
   A. she had proven to be an unfit mother
   B. she had signed a contract that was legally binding
   C. she was not the genetic mother of the child
   D. the rights of the child's genetic father were given priority

15. At the time of the Roe v. Wade decision, how many states provided for unconditional *abortion on demand*? 15.____
   A. None   B. Four
   C. Twenty-three   D. Forty-two

16. During the 1988 presidential campaign, candidate Gary Hart was caught in an extramarital affair which destroyed his candidacy. 16.____
The majority of American women who responded to this affair judged Hart's GREATEST fault to be

A. stupidity for allowing himself to be caught
B. personal naivete
C. dishonesty and deception in his personal affairs
D. lack of seriousness about his political purpose

17. Each of the following assumptions about AIDS is a misperception EXCEPT: 17.___
A. There is an affinity between AIDS and certain sex practices
B. Certain sex practices cause the AIDS disease
C. Certain types of sex practices involve a greater AIDS infection risk than others
D. Some types of sex are symptomatic of the AIDS disease

18. During the Iran-Contra scandal, why did Oliver North become a popular figure among many Americans? 18.___
A. There was widespread support for the Nicaraguan Contra movement.
B. They didn't believe North had lied about his involvement in the affair.
C. There was no evidence that North had attempted to cover up the situation.
D. They believed North's lies to be patriotic service of the nation's interests.

19. The First Amendment 19.___
A. permits Holocaust revisionists the right to speak at a university
B. allows a false cry of *fire* in a crowded theater
C. allows Louis Farrakhan to address B'Nai Brith
D. prohibits the United States Congress from making laws abridging free speech

20. What kind of behavior is MOST widely tolerated by members of the business community because of the widespread belief that *everyone does it*? 20.___
A. Price gouging
B. Taking advantage of *insider* tips
C. Discriminatory hiring practices
D. Fixing prices among agreeing competitors

21. In the late 1980's, it was discovered that the reason for televangelist Jimmy Swaggart's staunch criticisms of promiscuity and homosexuality were PRIMARILY due to 21.___
A. financial backing from various conservative interest groups
B. a strongly literal interpretation of the Scriptures
C. self-loathing and doubt created by his own sexual obsessions
D. strong spiritual convictions about God and the nature of love

22. Public opinion of unethical behavior by prominent political figures is generally driven by which consideration among American citizens? 22.___
   A. Potential harm to individuals like themselves
   B. Gravity of the offense according to the law
   C. Gravity of the offense according to a personal code of ethics
   D. Potential harm to traditionally downtrodden groups

23. Why do many journalists claim that the ethic of objectivity is an unattainable ideal? 23.___
   A. Newspapers and television stations have commercial roots that cannot be severed.
   B. Human beings are not objective by nature.
   C. The public's interests and desires cannot be discounted.
   D. It is virtually impossible for a person to impartially observe and record facts.

24. Which of the following is a possible benefit of television as a source of political campaign information? 24.___
   A. It is less likely to be influenced by corporate interference than other sources.
   B. Inconsistent statements by candidates are virtually impossible to conceal.
   C. It makes candidates wary about embarking on a negative, opponent-centered campaign.
   D. It presents more opportunities for in-depth discussion of platforms.

25. Why did Karen Quinlan, who died in 1985, spend ten years in a persistent vegetative state despite her family's desire to end her life? 25.___
   A. Her family could not legally have her respirator shut off.
   B. She was found to be capable of conscious thought.
   C. The courts decided that disconnecting her feeding tube would be illegal and cruel.
   D. She had signed a statement professing her desire to have her life prolonged by any means.

KEY (CORRECT ANSWERS)

1. D	6. B	11. B	16. C	21. C
2. D	7. D	12. C	17. C	22. A
3. A	8. C	13. D	18. D	23. A
4. B	9. A	14. B	19. D	24. B
5. A	10. D	15. B	20. B	25. C

TEST 2

DIRECTIONS: Each question or incomplete statement is followed by several suggested answers or completions. Select the one that BEST answers the question or completes the statement. *PRINT THE LETTER OF THE CORRECT ANSWER IN THE SPACE AT THE RIGHT.*

1. In the 1964 presidential election, Lyndon Johnson campaigned as a candidate who would, if elected, work for peace and the end of the Vietnam War, though he fully intended to escalate military activity upon election. What is the specific argument that provides an explanation why this is now considered unethical?
It
   A. was a brazen political lie
   B. contributed to an already immoral military exercise
   C. denied the public a chance to give or refuse consent for escalation
   D. placed the lives of thousands of American troops in jeopardy

   1.___

2. Which aspect of Bill Clinton's personal association with an Arkansas savings and loan organization caused the GREATEST concern among the American media and public during the Whitewater affair?
   A. The organization contributed to the nationwide S&L crisis during the 1980's.
   B. The organization pulled him into an unsuccessful real estate investment.
   C. A government loan had been granted the organization while he was governor.
   D. The organization discriminated against clients according to their political affiliations.

   2.___

3. Janet Cooke's 1981 Pulitzer Prize was won for an *investigative* book about an 8-year-old heroin addict which was later proven to be a work of fiction.
Her actions primarily violated which element of journalism's informal code of ethics?
   A. Maintaining balance   B. Avoiding harm
   C. Reflection on purpose   D. Avoiding bias

   3.___

4. The increasing social trend in the United States is for citizens to define themselves in terms of their
   A. individual goals and interests
   B. familial ancestry
   C. association with cultural or political groups
   D. common identity as Americans

   4.___

5. Most American feminists identify their criteria for differentiating between erotica and pornography as 5.___
   A. homosexual/heterosexual content
   B. the appearance of male participants
   C. ratio of *hard* to *soft* content
   D. the degree of degradation against female participants

6. Which of the following methods of altering images in political advertisements is considered to be unethical in virtually any situation? 6.___
   A. Editing techniques
   B. Special effects
   C. Computerized alteration techniques
   D. Visual imagery

7. Most opponents of campus speech codes base their arguments on the grounds that 7.___
   A. the codes are more likely to work against members of certain ethnic groups
   B. most of the codes are arbitrarily constructed to remedy specific situations
   C. the codes are not specific enough to be enforced
   D. the codes violate the Constitution's free speech amendment

8. During the 1960's-70's, retarded children at the Willowbrook State School in New York City were subjected to a medical study that was later found to violate the *informed consent* element of the medical code because researchers 8.___
   A. made payments to the parents of the subjects
   B. did not study a large enough population to receive a representative sampling
   C. knowingly exposed the children to hepatitis during the study
   D. carried out the study without the knowledge of the community

9. After the 1988 presidential election, George Bush broke his *No New Taxes* campaign pledge by agreeing to the largest single tax increase in American history before Clinton's tax increase of 1993. 9.___
   The breaking of his promise initially resulted in very little criticism from the American public because 70% of the people who voted for Bush
   A. were unaware of his campaign pledge
   B. wanted him to raise taxes
   C. were unaffected by the tax increase
   D. didn't believe his pledge but voted for him anyway

10. The Savings and Loan crisis, which resulted in the loss of millions of tax dollars, was discovered by the news media ten years before widespread financial collapse made it a national crisis. 10.___
    How do media analysts explain the public's ten-year ignorance of the situation?

A. The public heard and read stories on the situation but were unable to comprehend the implications.
B. Despite widespread coverage, it was not considered to be an important issue until it became a crisis.
C. It received little coverage because it was considered too complicated for the public to understand.
D. It was covered up by prominent legislators who were invested in the savings and loan industry.

11. Until the late twentieth century, the sexual activity of American women was influenced by the Freudian belief that women who achieved orgasm solely through clitoral stimulation 11.___
A. were likely to be victims of past sexual abuse
B. were in an arrested state of psychological girlhood
C. tended to be subconsciously lesbian in partner preference
D. were psychologically more mature than other women

12. Currently, the LARGEST portion of a typical American candidate's political expenditures go toward 12.___
A. campaign staff payroll
B. contributions to interest groups
C. travel expenses
D. production of television commercials

13. What was the PRIMARY legal argument used by the defendants on trial for the 1992 beating of Reginald Denny during the Los Angeles riots? 13.___
They
A. were victims of a *riot* mentality that was allowed to proliferate by inept municipal authorities
B. acted in self-defense against a perceived threat to their physical well-being
C. were provoked by Denny's belligerent behavior
D. were innocent of striking or assaulting Denny

14. Generally, what is the criterion for judging the degree to which a political or financial conflict of interest is unethical? 14.___
A. The degree to which each competing interest profits
B. The balance of one's pursuit of both interests
C. Whether one of the interests is associated with the government
D. Creation of an unjustified effect on one's work

15. Under current federal laws, first trimester abortions are permissible 15.___
A. under any circumstances
B. only if the fetus is determined to be deformed or abnormal
C. only if the mother's health is threatened
D. only if the mother or parents can prove insolvency

16. During Justice Clarence Thomas' Supreme Court nomination hearings, what was NOT a reason often given by supporters of Anita Hill for their belief in the credibility of her sexual harassment charges? 16.___
   A. Her political affiliation
   B. Her initial reluctance to come forth with testimony
   C. The consistency of her testimony
   D. Her choice not to file a civil lawsuit against him

17. A *conflict of obligation* occurs in business or politics when an individual 17.___
   A. fails to commit to an obligation that is in the public interest
   B. is able to meet fewer obligations than he has committed himself to
   C. fails to meet his obligation to a constituent or client
   D. commits to more than one obligation

18. In the American legal arena, the attorney-client confidentiality privilege 18.___
   A. is a universally accepted right
   B. is an optional allowance in civil cases
   C. does not apply to knowledge of future crimes
   D. is nullified once a client admits to illegal activity

19. Which type of behavior by an employee constitutes a conflict of interest that can result in termination by some corporations? 19.___
   A. Soliciting information about employment from a competitor
   B. Affiliation with government agencies that could regulate the product's marketing and development
   C. Proven use of a competitor's product
   D. Apparent lack of enthusiasm for company products

20. Which of the following is the MOST effective deterrent to unethical behavior by politicians or public figures? 20.___
   A. Laws and regulations
   B. Voluntary codes
   C. News media scrutiny
   D. Popular judgment or backlash

21. A prevalent claim used by critics of Jack Kevorkian's assisted suicides is that they 21.___
   A. violate the doctor's Hippocratic oath
   B. use cruel and painful methods for ending life
   C. violate certain federal laws that forbid such actions
   D. are administered without informed consent of the patients

22. What, generally, has been the MOST prominent effect of the AIDS epidemic in the American social arena? 22.____
   A. Increased empathy for the homosexual lifestyle
   B. A marked proliferation of monogamous relationships and marriage
   C. A measurable decrease in homosexual activity
   D. A tempering of attitudes engendered during the *sexual revolution* of the 1960's-70's

23. In the mid-1980's, the nature of televangelist Oral Roberts' appeal to his followers for financial contributions sparked controversy because he 23.____
   A. claimed false financial hardships
   B. claimed God had told him he would be *called home* (die) if too little money was contributed
   C. was found to be speculating in risky financial ventures that were losing him millions of dollars
   D. stated the appeal by questioning each follower's loyalty to his church

24. The MINIMUM criterion required in order to justify the termination of a worker's employment is 24.____
   A. poor job performance
   B. personal incompatibility with other workers or management
   C. conflict of interest
   D. companies are not required to provide a reason

25. In 1992, the unethical practice of *staging* news events occurred at which level of reporting during the General Motors crash-test scandal? 25.____
   A. National news periodical
   B. Metropolitan news radio
   C. Regional news broadcast
   D. National network television

KEY (CORRECT ANSWERS)

1. C
2. C
3. C
4. C
5. D

6. C
7. D
8. C
9. D
10. C

11. B
12. D
13. A
14. D
15. A

16. C
17. B
18. C
19. C
20. D

21. A
22. D
23. B
24. D
25. D

TEST 3

DIRECTIONS: Each question or incomplete statement is followed by several suggested answers or completions. Select the one that BEST answers the question or completes the statement. *PRINT THE LETTER OF THE CORRECT ANSWER IN THE SPACE AT THE RIGHT.*

1. Which of the following is NOT an increasing trend in the American broadcast media? 1.___
   A. More channels
   B. Greater regulation of propagandist content
   C. Smaller audiences per channel
   D. More specific programming

2. In 1990, Nancy Cruzan had lain in a persistent vegetative state for eight years, without the aid of a respirator but sustained by a feeding tube. 2.___
   The courts decided to allow her family to disconnect the tube because
   A. a feeding tube was decided to be medical treatment that could be willfully suspended
   B. of an unfavorable cost/benefit analysis
   C. a family's decision was ruled to take precedence over all medical opinions
   D. the time passed was considered to be a long enough trial period for recovery

3. During the 1988 presidential campaign, candidate Joseph Biden was found to have been guilty of several ethical violations, the MOST serious of which was 3.___
   A. lying about his ancestry
   B. plagiarizing some of his law school work
   C. lying about his accomplishments
   D. plagiarizing parts of his speeches from several prominent historical figures

4. Of the following arguments against homosexuality, the one that has been MOST successfully disproven is that it 4.___
   A. is a biological anomaly in the animal kingdom
   B. involves exclusively recreational sex
   C. is forbidden by the Christian Bible
   D. is a learned behavior that can be changed

5. The primary grounds for the move to divest from apartheid-era South Africa were that 5.___
   A. any nation that practiced racial discrimination should not be allowed to benefit from the American economy
   B. the country's elections did not allow black candidates
   C. South Africa's aggressive and oppressive foreign policy
   D. almost all of the money invested was never received by members of the nation's majority population

6. The practice of *selective nontreatment* for infants born with physical defects, who later die from nontreatment, is sometimes legally exercised by American parents of children who 6.___
   A. are likely to place a large financial burden on the parents
   B. also suffer from accompanying genetic defects such as Down's syndrome
   C. are unable to be placed with an adoption agency
   D. could not survive even if treatment were administered

7. Former President Jimmy Carter has been largely judged by historians and the media to have been a failed president because 7.___
   A. he was not intellectually fit for the job
   B. his administration was among the most corrupt in American history
   C. his cabinet was made up primarily of personal friends, rather than qualified candidates
   D. his personal integrity and honesty prevented him from making the proper political decisions

8. Primarily, the sexual revolution of the 1960's-70's resulted in the perceived encouragement of which type of behavior? 8.___
   A. Homosexuality
   B. Promiscuity
   C. Pedophilia
   D. Aberrations such as sado-masochism

9. Currently, what is the APPROXIMATE annual abortion rate in the United States? 9.___
   A. 200,000
   B. 750,000
   C. 1.6 million
   D. 3 million

10. In the 1990's, the parents of Anissa Ayala aggressively pursued the conception of a second child whose primary purpose was to serve as a bone-marrow donor for their daughter, who was dying of cancer. 10.___
    This decision was eventually accepted by the public for each of the following reasons EXCEPT:
    A. The Ayalas claimed they would not abort the fetus if it proved to be an incompatible donor
    B. There was little hope for Anissa's survival without a suitable donor
    C. The Ayalas were shown to be loving and capable parents
    D. There was no legal precedent to decide the validity of the Ayalas' decision

11. An increasing trend among feminist lesbians in America is 11.___
    A. considering lesbianism to be primarily a political definition of a way of life
    B. moving away from social involvement and contact with males

C. active recruitment of younger women into the lesbian lifestyle
D. considering lesbianism to be an exclusively sexual definition of a way of life

12. In 1989, Larry McAfee, who had become a quadriplegic in a motorcycle crash, was granted permission to shut off a ventilator that was sustaining his life. 12.___
The legal rationale for this was
A. assisted suicides were becoming common practice
B. an individual's right to die took precedence over all medical considerations
C. McAfee was able to prove the pain and worthlessness of his daily routine
D. other patients who had not been declared brain-dead were granted the right to die by the same court

13. What is considered to be the MOST tangible result of the sexual revolution of the 1960's-70's? 13.___
A. A decline in the American marriage rate
B. A decreasing attachment to traditional family values
C. An increase in homosexual activity
D. The *commercialization* of sex and sex products

14. Which of the following elements of a hypothetical federal campaign ethics commission is NOT broadly supported by elected legislators? 14.___
A. Power to levy fines and penalties
B. Membership that is elected by the public
C. A budget for publicizing abuses
D. The power to order *corrective* ads that will remedy lies and smears

15. Currently, what APPROXIMATE percentage of Americans support the abortion of abnormal fetuses (fetuses with physical or mental defects)? 15.___
A. 10% B. 30% C. 70% D. 95%

16. During the 1988 presidential campaign, candidate Gary Hart was caught in an extramarital affair which destroyed his candidacy. 16.___
The majority of American men who responded to this affair judged Hart's GREATEST fault to be
A. stupidity for allowing himself to be caught
B. personal naivete
C. dishonesty and deception in his personal affairs
D. lack of seriousness about his political purpose

17. The downfall of television evangelist Jim Bakker was caused by a number of ethical violations. 17.___
Which caused the GREATEST loss of support from his followers?
A. His commission of adultery
B. His use of contributors' money for extravagant personal luxuries
C. Unethical financial speculation
D. The attempted blackmail of incriminating witnesses

18. Of the following public figures involved in the Iran-Contra scandal, who is the only one proven to have been unwilling to participate in the cover-up? 18.___
   A. George Bush
   B. Richard Secord
   C. John Poindexter
   D. Oliver North

19. Paula Corbin-Jones' sexual harassment charges against Bill Clinton were doubted by many members of the public and the media. 19.___
Which of the following was NOT generally a reason for this?
   A. Allegations against her by several close family members
   B. The high monetary amount sought by her lawsuit
   C. Widespread disbelief that the President was capable of acting with such impropriety
   D. The behavior described in her testimony did not legally amount to harassment

20. Anna Johnson, an American surrogate mother who carried a child fertilized *in vitro* with the sperm and egg of a married couple, was unable to secure visitation rights for the child because 20.___
   A. she was found to be mentally unsound
   B. she was not the child's genetic mother
   C. her contract did not include a clause addressing the question of visitation
   D. she had previously signed these rights away

21. Which element of the Iran-Contra scandal of the Reagan administration was met with the GREATEST amount of public disapproval? 21.___
The
   A. concessions made to terrorists
   B. potential danger to American hostages
   C. support of Nicaragua's Contra movement
   D. covertness of the operation

22. Prior to 1993, what degree of behavior was prosecutable under federal laws concerning sexual harassment in the workplace? 22.___
   A. Behavior that causes severe debilitation or impairment of the plaintiff
   B. Any behavior that is sexual in nature
   C. Behavior that is physically threatening to the plaintiff
   D. Any behavior that is offensive to the plaintiff

23. Which of the following was NOT a legal practice during the American eugenics movement of the early twentieth century? 23.___
   A. Compulsory sterilization of mothers in the third generation of low-IQ lineages
   B. Immigration restrictions based on IQ scores
   C. Deportation of immigrants with a low IQ
   D. Screening of immigrants according to racial backgrounds

24. In the early 1990's, a large percentage of Americans believed the Menendez brothers, Lyle and Erik, should not have been convicted of murdering their parents because they felt 24.___
   A. that alleged past abuse by the boys' father justified the murder
   B. there was not enough evidence to prove their guilt
   C. the brothers were innocent of the murder
   D. the brothers acted on impulse in self-defense

25. What is the current status of the *abortion pill*, RU-486, in the United States? 25.___
   A. It is illegal to prescribe or ingest it under any circumstances.
   B. Its approval for abortion indications is pending under FDA considerations.
   C. It is available for prescription as a birth control medication.
   D. It is legal only to induce first trimester abortions.

KEY (CORRECT ANSWERS)

1. B
2. A
3. D
4. D
5. D

6. B
7. D
8. B
9. C
10. D

11. A
12. C
13. D
14. B
15. C

16. A
17. A
18. B
19. C
20. B

21. D
22. A
23. C
24. A
25. B

TEST 4

DIRECTIONS: Each question or incomplete statement is followed by several suggested answers or completions. Select the one that BEST answers the question or completes the statement. *PRINT THE LETTER OF THE CORRECT ANSWER IN THE SPACE AT THE RIGHT.*

1. The majority of Americans tend to judge political candidates' competence PRIMARILY according to 1.___
   A. their ability to remain free of ethical accusations
   B. their responses to attacks of their ethics after charges are made
   C. the platform on which they run their campaigns
   D. their work record in the political arena

2. President Bill Clinton's difficulty in finding an acceptable candidate for the position of Attorney General after his 1992 inauguration was due to the failure of his first two candidates to pay the required tax for one or more of their personal employees. 2.___
   What was the PRIMARY reason why the public was opposed to the confirmation of these candidates?
   A. General resentment of those who do not pay taxes like the rest of the country's citizens
   B. Sympathy for exploited personal servants by American aristocrats
   C. Unconditional rejection of unethical behavior by a political figure
   D. Disapproval of the candidates' past performance as attorneys

3. Which of the following is NOT generally considered to be an ethical problem posed by the use of television in American political campaigns? 3.___
   The
   A. predisposition toward the dramatic and visual
   B. ability to influence virtually all members of American society
   C. exclusively high cost of buying broadcast time
   D. seeming ability of technology to alter reality

4. In 1991, Dr. Timothy Quill assisted a terminally ill patient who wanted to commit suicide by providing the patient with a prescription for barbiturates, on which the patient intentionally overdosed and died. 4.___
   This suicide assistance
   A. was broadly denounced by the public
   B. caused little controversy because Dr. Quill's participation was indirect
   C. was ruled to violate established state laws
   D. did not violate the spirit of the doctor's Hippocratic oath

5. Public opinion at the time of the Watergate affair showed that other than Watergate, the MOST serious of Nixon's ethical offenses throughout his political career was 5.___
   A. false and vicious campaign smears during his 1952 congressional campaign
   B. breaking into a law professor's office to change an undesirable academic grade
   C. the expressly forbidden invasion of Cambodia during the Vietnam War
   D. the use of tax dollars for improvements on his San Clemente home

6. Which of the following was NOT a crime associated with the Watergate scandal during the Nixon presidency? 6.___
   A. Perjury
   B. Burglary
   C. Blackmail
   D. Assault

7. American heterosexual relationships typically involve what sort of pressure on women that is not shared by their male partners? 7.___
   A. Demonstrating affection and tenderness
   B. Demonstrating physical stamina
   C. Pleasing one's partner
   D. Achieving orgasm

8. In the 1970's, what was the Ford Motor Company's justification for refusing to redesign the fuel tank structure of the Pinto, which was reputed to have caused over a hundred deaths? 8.___
   A. There was not enough solid evidence that the deaths had occurred because of poor design.
   B. The projected costs of the redesigning operation outweighed the potential benefits.
   C. There were very few complaints filed against the company.
   D. The fuel tank had performed well in several crash tests.

9. The 1972 Roe v. Wade Supreme Court decision, which grants a mother the right to abort a fetus under varying conditions, based its argument on which principle? 9.___
   A. The personal liberty provision of the 14th Amendment
   B. Established scientific opinion about the beginning of life
   C. The limited powers of states to make laws
   D. The freedom of expression provision of the 1st Amendment

10. Most government conflict-of-interest offenses are felonies according to the General Accounting Office, the Federal government's ethical watchdog.
Why do so few of the GAO's cases result in conviction? 10.___
   A. Most cases are difficult for prosecutors to prove.
   B. Juries are reluctant to return felony convictions against the defendants.

C. The GAO itself is suspected of being a corrupt bureau.
D. Most of the cases prove to be trumped-up partisan charges with little supporting evidence.

11. Under certain federal laws, a company may prevent an employee from accepting a position with a competitor 11.___
A. if the employee has not served the company adequate advance notice
B. if it can prove the employee is an unproductive worker
C. for as long as the company is unable to fill the position left behind by the former employee
D. if the employee is privy to trade secrets that would affect the competitive balance

12. During the Iran-Contra hearings of the late 1980's, Oliver North's testimony was considered by the media and the public to be truthful about 12.___
A. which country's government had conceived of the arrangement
B. his personal conviction that the situation was just and served the national interest
C. virtually all aspects of the conception and execution of the plan
D. the levels of the American government at which planning and consulting took place

13. In 1980, General Westmoreland's libel suit against CBS's 60 MINUTES was filed because the program, which accused Westmoreland of unethical behavior during the Vietnam War, 13.___
A. made unsubstantiated claims about his behavior
B. omitted testimony that called the program's claims into question
C. denied Westmoreland the right to defend himself on the program
D. included testimony from sources who were not qualified to make judgments about the situation

14. The 1932 Tuskegee Syphilis Study was recently denounced because its researchers violated the code of informed consent by 14.___
A. failing to inform subjects that they were being used for the study
B. infecting unknowing subjects with the disease
C. denying treatment to a number of *control* subjects, even though a cure was available
D. manufacturing data to support their hypothesis

15. The Supreme Court's 1972 Roe v. Wade decision 15.___
A. guarantees a woman's right to an abortion at any phase of pregnancy
B. disallows any state regulation of abortion funding
C. is designed to gradually cede jurisdiction to the community as a woman's pregnancy advances
D. is designed to maintain maximum individual jurisdiction during a woman's entire pregnancy

16. Of the information and debate that the public receives during a political campaign, what APPROXIMATE percentage comes from television? 16.___
A. 25% B. 50% C. 75% D. 100%

17. Which of the following is NOT an operation proven to have been carried out by the Federal Bureau of Investigation under the leadership of J. Edgar Hoover? 17.___
A. The maintenance of an active file of tapes and documents concerning the personal conduct of John F. Kennedy
B. The mailing of an anonymous letter to Martin Luther King, Jr.,urging him to commit suicide
C. Plotting to assassinate one or more Communist leaders in Southeast Asia
D. The spread of fictionalized rumors about Eleanor Roosevelt's sexual activity

18. When did the Federal government pass its first law creating an office to set standards for and investigate the conduct of government officials? 18.___
A. 1865 B. 1923 C. 1969 D. 1978

19. Which activity has the Central Intelligence Agency of the United States NOT been proven to have participated in? 19.___
A. Cooperating in shared pursuits with prominent Mafia leaders
B. Developing a plot to assassinate Fidel Castro, leader of Cuba
C. Conspiring to hinder the Warren Commission's investigation into John F. Kennedy's assassination
D. Gathering intelligence concerning other government agencies from personnel doubling as CIA agents

20. The American lawyer who permanently justified the circumstance of *lawyering for the situation*, or serving clients with conflicting interests under certain circumstances, was 20.___
A. William Jennings Bryan B. Clarence Darrow
C. Oliver Wendell Holmes D. Louis Brandeis

21. The McFall-Shrimp case in America concerned a man who was the only available and compatible bone marrow donor for a cousin who was dying of cancer, but who refused to submit to the donation procedure, a relatively simple and risk-free operation. 21.___
When the man's cousin sued to compel donation, the court ruled that
A. it was illegal for a suitable donor to deny another person's right to treatment under any circumstances
B. the cousin could not be compelled to undergo the donation procedure against his will
C. given the low-risk nature of the operation, the cousin would have to submit to the procedure
D. the cousin would have to submit only if another suitable donor could not be found

22. Currently, mandatory drug testing is most common and most legally sound among which segment of the American population? 22.___
   A. Participants in college athletics
   B. Applicants for government employment
   C. Applicants for employment with private companies
   D. Petitioners for government aid

23. The Beech-Nut Corporation was fined $2 million for labeling and marketing a sugar solution as *apple juice*. When did specific truth-in-labeling legislation pass at the Federal level in response to this and other incidents? 23.___
   A. 1972 B. 1980 C. 1989 D. 1994

24. In 1991, Antioch College's code of sexual consent mandated spoken consent for 24.___
   A. any type of activity that involves the potential exchange of bodily fluids
   B. each instance of physical contact between potential romantic partners
   C. sexual intercourse
   D. the vocalization of any phrase or comment that could be interpreted as sexual in nature

25. Under current law, a lawyer has a right to collect a fee for advice given to a client unless the 25.___
   A. advice does not produce favorable results
   B. client chooses not to follow the advice
   C. client has sought out a second opinion
   D. advice can be proven to be worthless

KEY (CORRECT ANSWERS)

1. B
2. A
3. B
4. B
5. D

6. D
7. D
8. B
9. A
10. B

11. D
12. B
13. B
14. C
15. C

16. C
17. C
18. D
19. C
20. D

21. B
22. C
23. D
24. B
25. D

EXAMINATION SECTION
TEST 1

DIRECTIONS: Each question or incomplete statement is followed by several suggested answers or completions. Select the one that BEST answers the question or completes the statement. *PRINT THE LETTER OF THE CORRECT ANSWER IN THE SPACE AT THE RIGHT.*

1. Ethnic identity as a factor in urban America 1.___
   A. has virtually disappeared with the rapid assimilation of second and third generation immigrants
   B. has little influence on patterns of occupational mobility
   C. has become an increasingly important determinant of residential choices
   D. continues to exercise an influence on voting behavior
   E. cannot be measured through sociological techniques

2. Advocates of the culture of poverty hypothesis maintain that remedial action should center on the 2.___
   A. discriminatory practices against minorities
   B. lack of work opportunity
   C. attitudes and behavior of the poor
   D. inequitable distribution of educational facilities
   E. redistribution of wealth by force

3. Perhaps the chief reason for the survival of many practices, customs, beliefs, and technological methods is 3.___
   A. stubbornness on the part of most people who say, *It was good enough for grandpappy, and it's good enough for me*
   B. pride
   C. selfishness
   D. that people are used to them
   E. the fear of *retribution from change*

4. Euthenics is the study of the methods of improving the quality of the population through 4.___
   A. improving the understanding of basic social needs
   B. applying knowledge of hereditary differences in reproduction
   C. elaborating the Aryan design and transferring its components to give a wider view of the benefits of endogamy
   D. the revival of euthanasia
   E. testing the effects of crossbreeding

5. In addition to habit, the fear of _____ often lies behind resistance to innovation. 5.___
   A. the unknown B. retribution C. retaliation
   D. God E. disruption

6. Many, perhaps most, sociologists agree that values, when they become organized into generalized ideologies, 6.___
   A. are the blocks over which new thoughts, new ideas, and inspired reasoning stumble
   B. completely inhibit social change
   C. like other factors in social change, are the result of still other changes, and they alone are the only pillars on which the advancement of the social process may stand
   D. usually degenerate into unquestioned dogma
   E. have an important influence on individual and collective behavior, hence, on social processes

7. In states where intermarriage between white people and people of Asian extraction is NOT forbidden, 7.___
   A. the laws of these states effectively prevent the non-whites from having a proper education, thereby making intermarriage undesirable
   B. there will be found a higher percentage of persons of Asian extraction
   C. there will be latent, underlying hostility, which makes such marriages practically impossible
   D. social disapproval often acts as the barrier
   E. only 15% of these marriages are celebrated publicly

8. Which of the following statements is an example of over-generalization in use among those who attribute crime to a single cause? 8.___
   A. The family that prays together, stays together.
   B. I knew he wouldn't amount to anything.
   C. All work and no play makes Al Capone a dull boy.
   D. The parents are to blame.
   E. Give 'em one inch and they'll want six.

9. Religion functions as a 9.___
   A. shield, not a sword
   B. cultural crutch
   C. constant source of sanction for the mores
   D. filter to prevent the dangerous tars and nicotines of truth
   E. justification for the individual's behavior

10. Among primitives, breaches of the laws of exogamy are 10.___
   A. frequent and ignored
   B. always punishable by death
   C. the most serious in their whole moral code
   D. not regarded as serious by the *elders* or leaders of the tribe but will result in complete ostracism of the violator by the younger members of the group
   E. punishable by death or banishment from the tribe forever

11. The _____ system is a very important determinant of social conduct in all primitive societies. 11.___
   A. capsian   B. Minoan   C. kinship
   D. dinaric   E. numinous

12. Moral rules are not always considered absolute. Sociologists believe that they derive their value from 12.___
   A. their use by the upper classes in imposing their will on the lower classes
   B. their great diversity and flexibility
   C. transcendental principles independent of human experience
   D. their social utility
   E. the ability of the clergy to *sell* them in a package deal with God

13. All primitive societies maintain a strict system of etiquette controlling one's behavior towards other people. Among the rules of etiquette, none is so important as that governing _____ relationships. 13.___
   A. maternal   B. impersonal   C. joking
   D. potlatch   E. business

14. Culture, or social heritage, surrounds social relationships in at least two ways. 14.___
It provides a set of values AND
   A. seeks to utilize mythological beliefs in justification of existence
   B. a set of norms
   C. sociofacts
   D. a sound platform on which modern man can stand
   E. attempts to show the way to idealized thinking

15. A change in customs leads to 15.___
   A. a change in mores
   B. a loosening of racial boundaries
   C. a decline of civilization in the broadest sense
   D. probable extinction of a sect or group
   E. the beginning of a *growing up* process

16. Wife-sharing has been practiced (and explained as a necessary pledge that a guest is a friend and not an enemy) by the 16.___
   A. Quakers   B. Eskimos   C. Amish
   D. Samoans   E. Tahitians

17. In a primitive society, one cannot marry whom one likes or anyone in one's own group; hence the term 17.___
   A. polyandry   B. exogamy   C. gynocracy
   D. matrilocal   E. patrilocal

18. Incestuous unions are the source of very powerful taboos; their origin and reason is 18.___
   A. a high rate of mortality
   B. biological

C. that negative or recessive genes are often involved
D. social
E. based upon religious beliefs

19. An important critic of American life says that in all the spheres where moral considerations are meant to operate, men act as though they were being guided by (or rebelling against) rules and prohibitions enunciated by 19.___
A. clergymen B. society C. a moral mother
D. Christ E. conscience

20. A normless society is termed 20.___
A. disorganized B. non-denominational
C. culturally lagging D. anomic
E. diffused

21. Sanctions are forms of reward or punishment based on conformity or 21.___
A. authoritarianism B. obedience
C. free enterprise D. non-conformity
E. rational conduct

22. In the ESSAY ON THE PRINCIPLE OF POPULATION, Malthus 22.___
A. advanced the idea that disease and war are necessary as the best possible checks on the population
B. supported birth control by contraceptive and other means
C. indicated that people would have to exercise voluntary restraints on reproduction
D. foresaw the day when the continuous succession of wars, famines, and plagues would end
E. realized that the food supply, broadly conceived to include other necessities, is also capable of great expansion with the development of new technologies and the more efficient division of labor

23. The study of _____ deals with the treatment of criminals by society. 23.___
A. criminology B. penology
C. multiple causation D. methodology
E. ecology

24. According to _____, sociology is built around the concepts of in-group and out-group, while folkways and mores are prescriptions for the treatment of members of one's own group as well as outsiders. 24.___
A. Ruth Benedict B. William Sumner
C. John Dewey D. Margaret Mead
E. Emile Durkheim

25. Two fundamentally opposed views clash in the treatment of the criminal, once convicted. On one hand is the position, embodied in the structure of criminal law, that the criminal must be punished because of his crimes against society and as a deterrent from further crimes. 25.___
On the other hand is the standpoint that

A. very few crimes justify the types of prison sentence, including capital punishment, meted out
B. trial by jury has been considered unjust because it is impossible to be judged by a jury of one's peers
C. the goal of treatment should be the protection of society and the rehabilitation of the criminal
D. prison life and the types of punishment imposed breed revenge in the criminal mind and often result in homosexuality
E. were paroles granted and good behavior rewarded more than it is at present, the criminal would react favorably towards society

KEY (CORRECT ANSWERS)

1. D
2. C
3. D
4. B
5. E
6. E
7. D
8. D
9. C
10. C
11. C
12. D
13. C
14. B
15. A
16. B
17. B
18. D
19. C
20. C
21. D
22. C
23. B
24. B
25. C

TEST 2

DIRECTIONS: Each question or incomplete statement is followed by several suggested answers or completions. Select the one that BEST answers the question or completes the statement. *PRINT THE LETTER OF THE CORRECT ANSWER IN THE SPACE AT THE RIGHT.*

1. Criminals are almost invariably 1.___
   A. a separate type of people
   B. without morals
   C. motivated by normal human desires
   D. *lone wolves* who do not seek or need the affection and approval of other people
   E. not found to subscribe to many of the mores to which law-abiding people adhere

2. Norms are supported by _____, which are forms of reward or punishment based on conformity or non-conformity. 2.___
   A. social attitudes
   B. sanctions
   C. logic and reason
   D. statutes and *unwritten laws*
   E. ethics

3. Since many norms are unconscious, _____ may arise when an individual is suddenly plunged into a different culture. 3.___
   A. chauvinism
   B. hypertension
   C. hysteresis
   D. xenophilia
   E. *cultural shock*

4. The *ethos* of a culture may be embodied in well-established customs of dress, diet, conformity to rituals, etc. Ethos is the sociological term for 4.___
   A. moral behavior
   B. blind conformity
   C. dominant spirit
   D. prejudices
   E. indoctrination

5. The eugenic movement favors 5.___
   A. a wider use of birth control among the poorer classes and a greater reproduction rate among the more fortunate
   B. birth control and euthanasia
   C. the restoration of sexual ability by the use of monkey glands
   D. the construction and operation of homes for the aged
   E. the rhythm method of birth control and the concept of original sin

6. The majority of people seem to practice monogamy, if only for _____ reasons. 6.___
   A. religious
   B. inconsequential
   C. sexual
   D. economic
   E. personal

7. Laws against miscegenation forbid 7.___
   A. the breeding of cattle
   B. marriage to foreigners
   C. intermarriage between Caucasians and members of other races
   D. the integration of school facilities
   E. anti-semitic practices

8. The process by which individuals or groups from one culture come to acquire the behavior and thought patterns of a different culture is termed 8.___
   A. osmosis B. intervention C. permeation
   D. correlation E. acculturation

9. The modern college has been compelled by circumstance to assume ____ its students. 9.___
   A. the financial support of
   B. a policy of complete integration of
   C. an *in loco parentis* attitude toward
   D. total academic responsibility for
   E. the additional role of the community with respect to

10. Nearly all prejudice is based on ____ thinking. 10.___
   A. deductive B. analytical C. inductive
   D. stereotyped E. intuitive

11. Which one of the following statements does NOT describe a function of group prejudice? 11.___
   It
   A. provides a source of egotistic satisfaction
   B. justifies various types of discrimination which are considered to be advantageous to the dominant group
   C. provides an outlet for aggressive feelings
   D. provides convenient scapegoats
   E. excludes most of those who are patently *undesirables* from participation in community activities

12. Class animosity does NOT increase in proportion to the ____ of social classes. 12.___
   A. competition
   B. differences in mores
   C. dominance and subjugation
   D. justifiable resentments
   E. separation

13. The medieval family was dominated by comparatively settled traditions which reflected the needs of the general system of society. 13.___
   Marriage was ESSENTIALLY regarded as a(n)
   A. result of love
   B. alliance of interests
   C. coming-of-age
   D. auction of the wife to the highest bidder
   E. means of ending inter-family feuds

14. As illustrated by the potlatch of the Kwakiutl Indians, the theory of _____ implies that, one shows, through extravagant waste, how much he has to throw away. 14.___
    A. diminishing returns
    B. superabundance
    C. *waste not, want not*
    D. capitalism
    E. conspicuous consumption

15. The use of human flesh as either a symbolic or regular food is often linked with ceremonials reflecting religious significance and is termed 15.___
    A. Catholicism
    B. vampirism
    C. cannibalism
    D. hara-kiri
    E. necrophilia

16. _____ comprehends both rights and duties. It is an individual's comparative prestige rank in a community. 16.___
    A. Status
    B. Caste
    C. Lineage
    D. Stratum
    E. Stasis

17. Ethnocentrism arises when a group 17.___
    A. departs from its environment
    B. fears attack
    C. intermarries
    D. feels superior culturally
    E. feels inferior

18. An established way of doing something which is recognized by the society as the *correct* way of doing it is a 18.___
    A. custom
    B. ritual
    C. positive reaction
    D. conformity
    E. non-creative act

19. Riesman's depiction of modern man as *other-directed* in contrast to *tradition-directed* and *inner-directed* leads to the widely-held view that modern man 19.___
    A. evaluates success in terms of the wealth he is able to amass
    B. evaluates success in terms of the approval of his peers
    C. is much too altruistic
    D. tends to discard all tradition regardless of its value
    E. regards other people as of relatively greater worth than himself

20. Shutting the door on travelers is almost unknown among 20.___
    A. Southerners
    B. preliterate cultures
    C. urbanites
    D. small businessmen
    E. American farmers

21. Among primitive peoples, _____ plays the LEAST important part in determining cultural roles. 21.___
    A. sex
    B. ability
    C. strength
    D. occupation
    E. age

22. Mores, according to Sumner, differ from folkways in that 22.___
   A. they are unchanging and universal
   B. they are religious in basis
   C. they result from a society's defense structure
   D. breaking them is considered a threat to a society's welfare, while violation of folkways can often occur with relative impunity
   E. the more different they appear, the more they are the same

23. Prescribed patterns of conduct which have the authority of, and are enforced by, the state are 23.___
   A. conventions B. customs C. mores
   D. folkways E. laws

24. Birth control and planned families are increasing 24.___
   A. among the Mormons
   B. among Catholics
   C. among rural folk
   D. among the clergy
   E. because of a serious food shortage in the Western Hemisphere

25. Total freedom is usually restricted by informal social pressures based upon _____, as well as by threats of retribution. 25.___
   A. religious doctrines B. the opinions of others
   C. government censorship D. statutes
   E. sound political theory

KEY (CORRECT ANSWERS)

1. C	11. E	21. D
2. B	12. E	22. D
3. E	13. B	23. E
4. C	14. E	24. B
5. A	15. C	25. B
6. D	16. A	
7. C	17. D	
8. E	18. A	
9. C	19. B	
10. D	20. B	

TEST 3

DIRECTIONS: Each question or incomplete statement is followed by several suggested answers or completions. Select the one that BEST answers the question or completes the statement. *PRINT THE LETTER OF THE CORRECT ANSWER IN THE SPACE AT THE RIGHT.*

1. The majority of marriages, past and present, throughout the entire world are 1.___
   A. polygamous B. monogamous C. polyandrous
   D. misogamous E. morganatic

2. The sociologist, in studying social problems, considers *values* as 2.___
   I. totally irrelevant
   II. important in the definition of a group-defined problem
   III. an integral element of social interaction
   IV. the subjective element of a social problem
   V. the objective element of a social problem

   The CORRECT answer is:
   A. I *only* B. II, V C. II, III
   D. II, III, IV E. II, III, V

3. The term *culture* denotes 3.___
   I. nothing in particular
   II. everything in general
   III. all the knowledge, values, beliefs, norms, and technology of a people
   IV. all the knowledge, norms, and technology in a people's history
   V. artistic achievements of a society

   The CORRECT answer is:
   A. I *only* B. II *only* C. V *only*
   D. II, III E. III *only*

4. To the sociologist, *values* are 4.___
   I. criteria of desirability
   II. vague and subjective
   III. objective and quantifiable
   IV. used as justifications for behavior
   V. unavoidable

   The CORRECT answer is:
   A. I, V B. II *only* C. III *only*
   D. I, IV, V E. I, III, V

5. American culture can be sociologically characterized by the values of 5.___
I. sloth and hedonism
II. achievement and work
III. efficiency and freedom
IV. progress and practical usefulness
V. self-awareness and growth

The CORRECT answer is:
A. I *only* B. V *only* C. II, V
D. II, III E. II, III, IV

6. A norm is considered *legitimate* when 6.___
I. an individual considers himself bound by it
II. society places punishments upon violations of that norm
III. everyone believes it
IV. sociologists determine its objective validity
V. it becomes someone's duty

The CORRECT answer is:
A. I, V B. II, III C. III *only*
D. IV *only* E. II, III, IV, V

7. A social institution is something that 7.___
I. is based on patterns of norms
II. can be observed directly
III. cannot be observed directly
IV. causes behavior which can be studied
V. causesrecurrent situations which can be studied

The CORRECT answer is:
A. I, II B. I, III C. I, IV
D. I, III, V E. II, IV

8. For the sociologist, marriage is considered to be 8.___
I. a matter of biological necessity
II. evolutionary activity
III. conducive to psychological health
IV. a mating arrangement involving norms of sex behavior
V. connected to responsibilities of adult mates

The CORRECT answer is:
A. I, II B. III *only* C. IV, V
D. I, IV E. I, III, V

9. A kinship system is 9.___
I. a set of statuses made of geneological relationships
II. a set of rights, privileges, and obligations
III. a system setting authorities
IV. relatively important in American social life
V. relatively unimportant in American social life

The CORRECT answer is:
A. I, II, IV B. I, II, III, IV C. I, II, V
D. I, II, III, V E. II, V

10. Sources of family discord include 10.___
I. the generation gap
II. longer periods of adolescence
III. vagueness attached to the role of adolescents
IV. changing mores
V. isolation of adolescents

The CORRECT answer is:
A. I, IV
B. II, III, V
C. II, IV
D. I, III, IV
E. All of the above

11. Social taboos in almost all societies include those on 11.___
I. murder
II. incest
III. adultery
IV. promiscuity
V. cannibalism

The CORRECT answer is:
A. I, II
B. II, III
C. II, V
D. II, IV, V
E. All of the above

12. *Polygyny* is a marriage institution in which 12.___
I. husbands may have more than one wife
II. wives may have more than one husband
III. husbands may have only one wife
IV. wives may have only one husband
V. both husbands and wives may have only one spouse

The CORRECT answer is:
A. I *only*
B. III, IV, V
C. IV *only*
D. II *only*
E. I, V

13. *Polygamy* is a marriage institution in which 13.___
I. husbands may have more than one wife
II. wives may have more than one husband
III. both husbands and wives may have more than one spouse
IV. wives may have only one husband
V. both husbands and wives may have only one spouse

The CORRECT answer is:
A. I *only*
B. I, III
C. III *only*
D. II, V
E. III, IV

14. *Polyandry* is a marriage institution in which 14.___
I. husbands may have more than one wife
II. wives may have more than one husband
III. both husbands and wives may have more than one spouse
IV. wives may have only one husband
V. both husbands and wives may have only one spouse

The CORRECT answer is:
A. I *only*
B. II *only*
C. III, IV
D. II, III, IV
E. V *only*

15. *Monogamy* is a marriage institution in which 15.___
   I. husbands may have more than one wife
   II. wives may have more than one husband
   III. both husbands and wives may have more than one spouse
   IV. wives may have only one husband
   V. both husbands and wives may have only one spouse

   The CORRECT answer is:
   A. I, III   B. II *only*   C. III, IV
   D. V *only*   E. IV *only*

16. Sociologically speaking, laws against some types of sexual behavior 16.___
   I. misplace social control as a substitute for morality
   II. place in the public realm what ought to be private
   III. enforce obligations rather than establishing norms
   IV. misunderstand social deviance
   V. are ill-worded and enforced whimsically

17. The term *hypergamy* is used to refer to marriage between a 17.___
   A. man and woman of the same caste
   B. man of high caste and a woman of low caste
   C. man of low caste and a woman of high caste
   D. Moslem and Hindu
   E. Hindu and Buddhist

18. In spite of the informality of the Norton Street Gang, 18.___
   A. it involved a set of mutual rights and obligations for all members
   B. it was capable of concerted action for neighborhood improvement
   C. the leader had absolute control and was entitled to unquestioning support
   D. status of members was determined solely by performance in objective tests, such as sports contests
   E. the leaders often wore tuxedos

19. Beliefs and values 19.___
   I. differ in that beliefs are ideas whereas values are desirable objects
   II. differ in that beliefs concern questions of truth whereas values concern questions of decision-making
   III. are both crescive institutions
   IV. are types of shared normative commitments

   The CORRECT answer is:
   A. I *only*   B. II, III
   C. II, IV   D. I, II
   E. All of the above

20. Differences in behavior, attributed by *racist* theories to differences in race, are MORE often attributable to differences in 20.___
   A. intelligence   B. language
   C. cultural expectations   D. population size
   E. age

21. Institutions are 21.___
    A. shared definitions of appropriate conduct
    B. formally organized permanent groups
    C. major requirements that must be fulfilled by any society if it is to survive
    D. customs which develop among people contrary to enacted laws
    E. academic think tanks

22. The variations in the incest taboo reflect 22.___
    A. racial differences
    B. differences in family structure
    C. religious differences
    D. class differences
    E. personal choices

23. *Personality* refers to the phenomenon in the individual's make-up of 23.___
    A. organization of habits, ideas, goals, etc. into a coherent system
    B. individual variations in temperament and magnetism
    C. outstanding and idiosyncratic patterns which deviate from cultural norms
    D. cultural creativity and voluntary behavior found only in social primates
    E. occupation, age, and ability to maintain meaningful relationships

24. The fundamental observation underlying the role and status concept is that 24.___
    A. people act in accordance with commonly held norms
    B. people are expected to act in different ways, depending on who or what they are
    C. people's actions are based on the power of others to reward or punish them
    D. the influence of an institution over an individual depends on approval of the other people who support the institution
    E. people are generally forced into their roles and status

25. In a primary group, hostile relationships 25.___
    A. cannot be tolerated
    B. arise only with non-group members
    C. are not subject to group patterns of control
    D. are often generated and regulated by group norms
    E. are generally encouraged

KEY (CORRECT ANSWERS)

1. B	6. A	11. C	16. C	21. A
2. D	7. D	12. A	17. B	22. B
3. E	8. C	13. C	18. A	23. A
4. D	9. D	14. B	19. D	24. B
5. E	10. E	15. D	20. C	25. D

TEST 4

DIRECTIONS: Each question or incomplete statement is followed by several suggested answers or completions. Select the one that BEST answers the question or completes the statement. *PRINT THE LETTER OF THE CORRECT ANSWER IN THE SPACE AT THE RIGHT.*

1. Personal feelings of guilt 1.____
   A. arise as psychological responses in the individual irrespective of the society in which he lives
   B. do not arise unless the individual is socialized to expect them
   C. are responses to actual or potential violations of social norms and are sometimes encouraged by social situations
   D. operate as psychological, but not social, forces
   E. are the result of parental pressure

2. American marriage patterns are distinguished by emphasis on 2.____
   A. having children
   B. legitimacy of sexual relations
   C. love
   D. compatibility of temperament
   E. fear of being alone

3. Uniform behavior can be expected of members of a social group despite individual differences since 3.____
   I. the members have been socialized to follow common norms
   II. group organization results in rewards for conformist behavior and penalties for deviant behavior
   III. admittance to the group is dependent on conformity to group norms
   IV. groups with nonconforming members tend to disintegrate

   The CORRECT answer is:
   A. I, III B. II, IV C. III, IV
   D. I, II E. II, III

4. Which one of the following choices BEST describes the tactic of non-violent resistance as used by civil rights groups? 4.____
   The
   A. willingness of persons to accept unlawful arrest without resistance
   B. avoiding of prosecution for violations of law by refusing to appear in court when required
   C. teasing and verbal harassment of police officers in order to cause unlawful arrests
   D. intentional violation of a particular law by persons unwilling to accept the penalty for violating that law
   E. intentional violation of a particular law by persons willing to accept the penalty for violating that law

5. Probably the MOST important single factor involved in the acceptance of birth control techniques is 5.___
   A. the economic situation
   B. religion
   C. education
   D. science
   E. availability

6. The strain facing a working-class youth in contact with middle-class values will exist 6.___
   A. regardless of his own values
   B. to the extent to which he accepts middle-class goals
   C. only in urban environments
   D. only in rural environments
   E. only in America

7. A public assistance program objective should be designed to 7.___
   A. provide for eligible persons in accordance with their individual requirements and with consideration of the circumstances in which they live
   B. provide for eligible persons at a standard of living equal to that enjoyed while they were self-supporting
   C. make sure that assistance payments from public funds are not too liberal
   D. guard against providing a better living for persons receiving aid than is enjoyed by the most frugal independent families
   E. eliminate the need for private welfare agencies

8. The scientific ethos rejects the value of 8.___
   A. universalism
   B. knowledge being private property
   C. communism
   D. skepticism
   E. cooperation

9. One basic cleavage within Protestantism centers around the question of 9.___
   A. religious intervention in political affairs
   B. support of capitalist practices
   C. the nature of its relationship to other religions
   D. national sovereignty
   E. the existence of an afterlife

10. Ideologies are composed of 10.___
   A. values
   B. interpretations of history
   C. views of the nature of man and society
   D. concepts on human life and culture
   E. all of the above

11. The super-ego is formed through the 11.___
   A. identification of the child with the parent and parental standards
   B. actions of the unconscious
   C. non-adoption of some societal norms
   D. processes of maturation
   E. adoption of erroneous ideas

12. In Merton's analysis, deviant behavior results from 12.___
    A. the discrepancy between cultural goals and opportunities to achieve them
    B. the desire to escape from cultural goals
    C. personality difficulties
    D. hereditary reasons
    E. lack of education

13. The fact that most forms of deviance are not distributed evenly through the various strata of society offers support to the contention that 13.___
    A. deviance is a purely lower-class phenomenon
    B. middle-class values tend to generate only certain types of deviance
    C. psychological explanations of deviance are worthless
    D. the form and distribution of deviance are related to men's roles, institutions, and values, and their interconnections
    E. deviance is a controllable phenomenon

14. According to David Riesman, the nineteenth-century American tended to be 14.___
    A. highly sensitive to the prevailing views and opinions
    B. largely driven by his own views and conscience
    C. highly attentive to tradition
    D. family-oriented
    E. dominated by European views and opinions

15. The relationship of modern science to contemporary values has been 15.___
    A. comparatively slight since science is value-neutral
    B. great in the economic area but slight in the artistic and political areas
    C. great in all areas of culture since the decline of theological control of ideas
    D. great since socially-approved values determine the findings of science
    E. comparatively slight since values are generally unscientific

16. An ideology may serve to 16.___
    A. distort societal realities
    B. sustain an opposition group
    C. justify and defend the cultural values
    D. unite otherwise dissimilar groups
    E. all of the above

17. The concept of *internalization* refers to the fact that 17.___
    A. the individual learns to accept the values and norms of his society
    B. cultural values and norms are imposed upon the individual in the process of socialization
    C. cultural values and norms come to form a part of the individual's personality

D. the individual never completely accepts the values and norms of his society
E. cultural values and norms are not externally visible

18. The notion of a tension between the fundamental drives of the individual and the requirements of living in society was expounded by 18.___
A. Durkheim B. Freud C. Malinowski
D. Kluckhohn E. Weber

19. Resentment differs from rebellion in that it 19.___
A. condemns traditional values
B. sanctions the current norms
C. involves an underlying commitment to traditional values
D. is a response to frustration
E. denies the existence of a conflict

20. Value conflicts may be resolved by 20.___
A. ignoring their existence
B. establishing a ranking of dominant and subordinate values
C. deviant behavior
D. expediency norms
E. fighting

21. The mass media have a dual role in the socialization process because children learn directly from them while 21.___
A. parents absorb from the mass media the standards of behavior which they transmit to their children
B. the media may be controlled by the government or by various interest groups
C. unintended consequences may result from mass media content and techniques
D. the media learns from the attitudes of children
E. all of the above

22. The importance of reciprocal obligations in ensuring compliance with social norms was shown by 22.___
A. Durkheim B. Freud C. Malinowski
D. Gouldner E. Weber

23. The school functions as an agency of socialization in part through 23.___
A. providing an opportunity for peer-group interaction
B. providing an opportunity for teacher-student relationships
C. providing an opportunity for the transmission of cultural values
D. providing a model of the larger society
E. all of the above

24. The concept of cultural pluralism is MOST actively opposed by 24.___
    A. the Amish
    B. supporters of black studies as a discipline
    C. supporters of bilingual education
    D. supporters of parochial schools
    E. all of the above

25. A certain boy is raised by parents who are concerned with status, social position, the *right* occupation, the *right* friends, the *right* neighborhood, etc. Social behavior plays a vital role in their lives, and their outlook with regard to rearing children can best be summed up by *children should be seen and not heard*. 25.___
Following are four descriptive terms their son might possibly be likely to use if he were asked to describe the *perfect boy*:
    I. Being polite
    II. Being a good companion
    III. Being clean
    IV. Being fun
Which one of the following choices MOST accurately classifies the above statements into those the boy is MOST likely to use when describing the *perfect boy* and those which he is LEAST likely to use?
    A. Most likely to use I and II and least likely to use III and IV
    B. Most likely to use I and III and least likely to use II and IV
    C. Most likely to use I, II, and III and least likely to use IV
    D. Most likely to use II and IV and least likely to use I and III
    E. Equally likely to use any of I, II, III, and IV

KEY (CORRECT ANSWERS)

1. C
2. C
3. D
4. E
5. C

6. B
7. A
8. B
9. A
10. E

11. A
12. A
13. D
14. B
15. C

16. E
17. C
18. C
19. D
20. B

21. A
22. C
23. E
24. A
25. B

ANSWER SHEET

TEST NO. ____________ PART ______ TITLE OF POSITION ____________________
(AS GIVEN IN EXAMINATION ANNOUNCEMENT - INCLUDE OPTION, IF ANY)

PLACE OF EXAMINATION ____________________ (CITY OR TOWN) ____________ (STATE) DATE ____________

RATING

USE THE SPECIAL PENCIL. MAKE GLOSSY BLACK MARKS.

1 A B C D E	26 A B C D E	51 A B C D E	76 A B C D E	101 A B C D E
2 A B C D E	27 A B C D E	52 A B C D E	77 A B C D E	102 A B C D E
3 A B C D E	28 A B C D E	53 A B C D E	78 A B C D E	103 A B C D E
4 A B C D E	29 A B C D E	54 A B C D E	79 A B C D E	104 A B C D E
5 A B C D E	30 A B C D E	55 A B C D E	80 A B C D E	105 A B C D E
6 A B C D E	31 A B C D E	56 A B C D E	81 A B C D E	106 A B C D E
7 A B C D E	32 A B C D E	57 A B C D E	82 A B C D E	107 A B C D E
8 A B C D E	33 A B C D E	58 A B C D E	83 A B C D E	108 A B C D E
9 A B C D E	34 A B C D E	59 A B C D E	84 A B C D E	109 A B C D E
10 A B C D E	35 A B C D E	60 A B C D E	85 A B C D E	110 A B C D E

Make only ONE mark for each answer. Additional and stray marks may be counted as mistakes. In making corrections, erase errors COMPLETELY.

11 A B C D E	36 A B C D E	61 A B C D E	86 A B C D E	111 A B C D E
12 A B C D E	37 A B C D E	62 A B C D E	87 A B C D E	112 A B C D E
13 A B C D E	38 A B C D E	63 A B C D E	88 A B C D E	113 A B C D E
14 A B C D E	39 A B C D E	64 A B C D E	89 A B C D E	114 A B C D E
15 A B C D E	40 A B C D E	65 A B C D E	90 A B C D E	115 A B C D E
16 A B C D E	41 A B C D E	66 A B C D E	91 A B C D E	116 A B C D E
17 A B C D E	42 A B C D E	67 A B C D E	92 A B C D E	117 A B C D E
18 A B C D E	43 A B C D E	68 A B C D E	93 A B C D E	118 A B C D E
19 A B C D E	44 A B C D E	69 A B C D E	94 A B C D E	119 A B C D E
20 A B C D E	45 A B C D E	70 A B C D E	95 A B C D E	120 A B C D E
21 A B C D E	46 A B C D E	71 A B C D E	96 A B C D E	121 A B C D E
22 A B C D E	47 A B C D E	72 A B C D E	97 A B C D E	122 A B C D E
23 A B C D E	48 A B C D E	73 A B C D E	98 A B C D E	123 A B C D E
24 A B C D E	49 A B C D E	74 A B C D E	99 A B C D E	124 A B C D E
25 A B C D E	50 A B C D E	75 A B C D E	100 A B C D E	125 A B C D E

ANSWER SHEET

TEST NO. ____________ PART ________ TITLE OF POSITION ______________________________
(AS GIVEN IN EXAMINATION ANNOUNCEMENT - INCLUDE OPTION, IF ANY)

PLACE OF EXAMINATION ______________________________ DATE __________
(CITY OR TOWN) (STATE)

RATING

USE THE SPECIAL PENCIL. MAKE GLOSSY BLACK MARKS.

1	A B C D E	26	A B C D E	51	A B C D E	76	A B C D E	101	A B C D E
2	A B C D E	27	A B C D E	52	A B C D E	77	A B C D E	102	A B C D E
3	A B C D E	28	A B C D E	53	A B C D E	78	A B C D E	103	A B C D E
4	A B C D E	29	A B C D E	54	A B C D E	79	A B C D E	104	A B C D E
5	A B C D E	30	A B C D E	55	A B C D E	80	A B C D E	105	A B C D E
6	A B C D E	31	A B C D E	56	A B C D E	81	A B C D E	106	A B C D E
7	A B C D E	32	A B C D E	57	A B C D E	82	A B C D E	107	A B C D E
8	A B C D E	33	A B C D E	58	A B C D E	83	A B C D E	108	A B C D E
9	A B C D E	34	A B C D E	59	A B C D E	84	A B C D E	109	A B C D E
10	A B C D E	35	A B C D E	60	A B C D E	85	A B C D E	110	A B C D E

Make only ONE mark for each answer. Additional and stray marks may be counted as mistakes. In making corrections, erase errors COMPLETELY.

11	A B C D E	36	A B C D E	61	A B C D E	86	A B C D E	111	A B C D E
12	A B C D E	37	A B C D E	62	A B C D E	87	A B C D E	112	A B C D E
13	A B C D E	38	A B C D E	63	A B C D E	88	A B C D E	113	A B C D E
14	A B C D E	39	A B C D E	64	A B C D E	89	A B C D E	114	A B C D E
15	A B C D E	40	A B C D E	65	A B C D E	90	A B C D E	115	A B C D E
16	A B C D E	41	A B C D E	66	A B C D E	91	A B C D E	116	A B C D E
17	A B C D E	42	A B C D E	67	A B C D E	92	A B C D E	117	A B C D E
18	A B C D E	43	A B C D E	68	A B C D E	93	A B C D E	118	A B C D E
19	A B C D E	44	A B C D E	69	A B C D E	94	A B C D E	119	A B C D E
20	A B C D E	45	A B C D E	70	A B C D E	95	A B C D E	120	A B C D E
21	A B C D E	46	A B C D E	71	A B C D E	96	A B C D E	121	A B C D E
22	A B C D E	47	A B C D E	72	A B C D E	97	A B C D E	122	A B C D E
23	A B C D E	48	A B C D E	73	A B C D E	98	A B C D E	123	A B C D E
24	A B C D E	49	A B C D E	74	A B C D E	99	A B C D E	124	A B C D E
25	A B C D E	50	A B C D E	75	A B C D E	100	A B C D E	125	A B C D E